Pyramid Geometry Design: Mathematics With Relatively Light Physics' Foundations

Timothy S. Margulies

2014

ISBN-13:978-1500445034

ISBN-10:1500445037

Preface

Theories of the ancient monuments design are visited to apply mathematics and the physical sciences to help illuminate on their beauty and form. The mathematical ideas and tools are emphasized to possibly supplement current curricula; however, several specific numerical values or visual sciences are investigated. The Golden proportion and Fibronacci series are derived due to their almost ubiquitous application in nature, as well as external and internal Egyptian pyramid designs. Additionally, the star light from constellations in space and planetary data are reviewed.

In particular, the architecture and physical sciences motivate navigating with mathematics.

Several topics addressed include,

Solid geometry of pyramids
Arithmetic-algebra for volume of pyramids calculation
Arithmetic sequence sum
Geometric progression sum
Quadratic equation solution for Golden proportion
Golden Section geometry construction
Chords and secant laws of a circle
Right triangle trigonometry and the solution of Euler's problem
Spherical trigonometry laws of cosines and sines
Equation for a plane in three dimensional space
Conic Section Ellipse, Hyperbola, and Parabola Curves by Construction
Determinant formulas for line, plane, triangle area, and fitting a second degree equation to data
A Numerical Estimation of Pi
Matrix applications (system of algebraic equations, optics, analytic hierarchy process)
Matrix linear regression applied to planetary orbit calculation
Statistical frequency histograms and average, median, and standard deviation statistical measures
Calculus integration to solve ordinary differential equation for elliptical motions of planets
Snell's Law of Refraction
Special relativity and color
Quantum mechanics wavelength Golden Phi ratio
General Relativity Hubble Universe Expansion-Contraction
Fractional Calculus Kinetics of Star Composition

Table of Contents

Conic Section Constructions
Geometry Theorems
Several Papyri Problems Revisited
Euler's Trigonometric Identity
Spherical Trigonometry
Analytic Hierarchy Preference Matrix
Transfer Matrices in Optical Elements
On Central Force Planet Motion Elliptical Orbits
Examples of Determinant Notation for Several Geometry and Algebra Problems
Special Relativity Principle for Light Waves
Modeling Expansion-Contraction of the Universe
Snell's Law and Atmospheric Refraction
Fractional Calculus Kinetics

Summary

The population of pyramids are surfacing new insights and corroborating earlier hypotheses; in particular, the Giza site arrangement of its major pyramids with Constellation brightness correlations to supplement the many Golden proportion ratios in pyramid measurements. Both the Orion's belt and Cygnus wing stars were fit by linear regression exhibiting the same slope trend of pyramid size and brightness. The frequency distribution of slope angles, or Egyptian Sekeds, volumes, and height-to base ratios from data by both Piazzi Smyth and Noel Wheeler are presented. Several mathematical formulas for volume of the truncated pyramid body are developed, noting similarities to several problems found in the Rhind and Moscow Papyri. To be in the light appears globally as exhibited by a robust order of pyramid monuments encircling the Earth which are statistically fit to a plane equation. Finally, a Nazca line drawing's interpretation depicts these markings of pyramids, light, and this global design project.

Introduction

Pyramids as physical and spiritual burial grounds to transform a body with ideas of soul embodiment of life energy from death to light from heavens above has received designers' attention to heavenly light sources, especially the Orion Constellation. Orion is the symbol for Osiris, a god idea of the underworld and resurrection. The three star *belt* or girdle refers to the Stars that were probably the aim as a narrow Star-gate to join with the Sun-pharaoh. The Cygnus Constellation which is another competitive theory for Star brightness and arrangement for the Giza complex is analyzed too .

An evolution of burial grounds from mastabas to stacked-mastabas as a stepped pyramids onto perfected ideas for the three kings at Giza which are regular square base pyramids ramping the king up in this ascension as a plausible explanation. The great pyramid Khufu, called Cheops in Greek, exhibited a shining whitish limestone casing from Turrah with good light reflection has become cream colored.

Earthly pyramid and site complex design, along with human form dimensions composed the quintessential forms of beauty expressed in right triangles, probably Golden Phi ratios, musical octaves, and scales. The first pyramid at Giza named after its owner Khufu Akhet (2589-2566 BCE) whose name means *Eastern Horizon*, the second Pharaoh Khafre Wer, or *Great is Khafre*, of the fourth dynasty, was son of King Seneferu and Hetepheres. The third pyramid is by Menkaure Netery meaning *Menkaure is Divine*. The perfected Great Giza Pyramid has been associated with meanings of both horizon and lights by Davidson as inspired architectural design.

The knowledge of arithmetic, geometry, and arithmetic-algebra appear as the foundational mathematics for these monuments. This eport first presents the mathematical ideas and formulas for finding volumes and Golden ratios of the pyramids, then presents measurement data and quantitative analysis where these mathematical ideas are embedded including a newer astroo-archaeo-anthropology investigations of light magnitude with psychometrics of vision concepts.

Pyramid Geometries and Volume

Geometry word origins mean *measure of the Earth*. The pyramid dimensions have been analyzed as incorporating Earth-Sun information, as well as practical construction. The square base is aligned with the cardinal points of the compass: North, South, East, and West with noticeable precision. The pramid is perpendicular to the strike slope of its limestone plateau. The angular bearing to Bethehem was recorded by Edgar and Davidson as the complement of the angle $26°18'10''$ which falls within the range for the slope angle of the Khufu Pyramid passages. Elaborate interpretation of the levels within the pyramid as records of history of wars and major events was revealed by Davidson with scripture passages as prophecy to an End Time during the mid 1930's. Wheeler comments that level accuracy was probably one to two centimeters beyond the hundreths of an inch accuracy of Edgar and Davidson's chronograph.

The linear and volume measures were important in the pyramid design. The derivations and formulas for the regular and oblique pyramid volumes calculations were probably specialty kowledge baes only known to few. However insights from the Golenishev, or Moscow, Papyrus named by a buyer in Thebes in approximately the year 1892, and the geometry of Euclid's *Elements* present several mathematical approaches and insights for volume calculation. Here several volume partitioned shapes are analyzed that form the pyramid volume calculation by arithmetic-algebra with a 1/3 fraction.

The first Egyptian pyramids were were a stepped design. For example, the Sakkarra Pyramid of the third dynasty was built approximately 2840 BCE. The base measurements are for a rectanglar base. Calculating a volume relies on an arithmetic summation such as found in the Rhind Papyrus.

Figure 1:   Five-Step Pyramid

T. Margulies, *Pyramid Geometry Design*

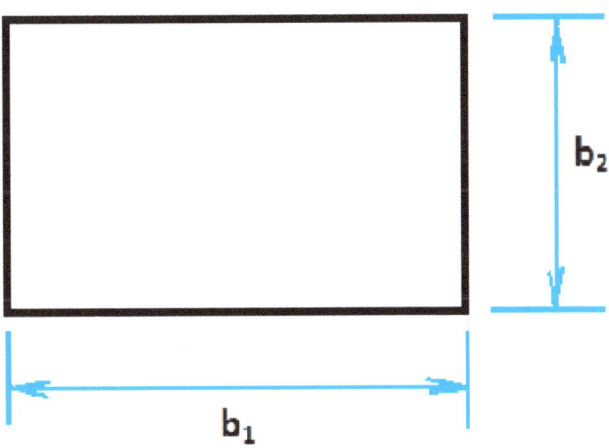

Figure 2: Rectangular Bases

Top base area, $B_1 = b_1 \cdot b_2$

Total sum of base areas, $S = B_1 + B_2 + B_3 + B_4 + B_5,$

$$B_2 = B_1 + d, B_3 = B_2 + 2d, B_4 = B_3 + 3d, B_5 = B_4 + 4d$$

Similar arithmetic summations were solved in Problem 64 with a different application of the papyrus.

$$V = \left[B_1 \cdot n + \frac{1}{2}n(n-1)d\right] h \ , \ \text{Here, n = 5}, \quad h = \frac{H}{n}$$

The mathematical formula for the volume of a regular pyramid is given by

$V = \frac{1}{3} Base \cdot Height$ presented using geometry ideas presented by Euclid, as well as the Moscow

Papyrus problem related to the truncated pyramid using arithmetic-algebra.

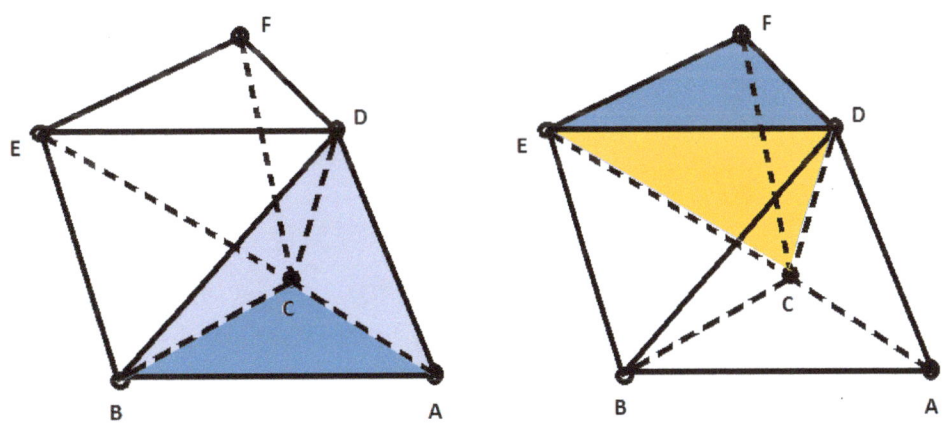

T. Margulies, *Pyramid Geometry Design*

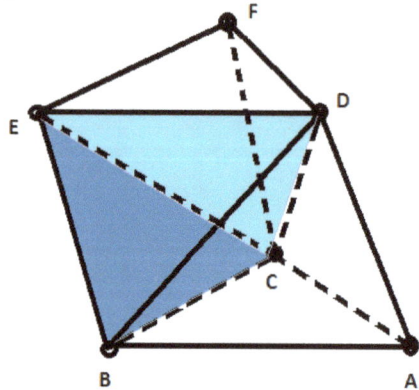

Figure 3

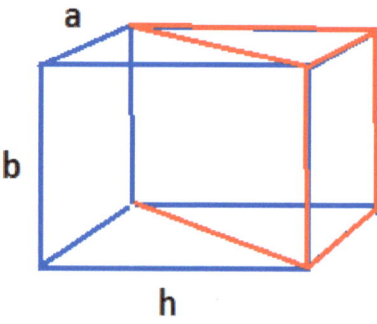

$$V = 0.5\,Bh, \quad V = \frac{1}{3}\,Bh\,, \; B = b \cdot b$$

Figure 4

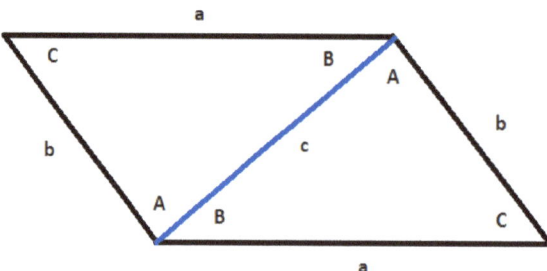

Figure 5

Euclid (325-265 BCE) is considered a father of geometry and writes proofs needed on the parallelogram and sold geometry prism or pyramid volumes. In particular, he proves the one-third factor used in the volume calculation for a pyramid with triangular base and apex vertically above one vertex of the base. This is the needed volume factor for both regular and oblique pyramid calculation. The three equivalent volumes of a prism ABCDEF are shown in Figure 3 for the oblique pyramid which relies on the parallelogram and similarity of triangle arguments. For example, ABED and FCBE with diameters BD and CE, respectively are parallelograms of congruent triangles. The proof finds that pyramid pairs 1 and 2, 3 and 4, 5 and 6, as well as 7 and 8, and 3 and 5 listed in the following table are equivalent pyramid constructions.

Table 1

Geometry pyramid Number	Base	Vertex
1	DEB	C
2	EBC	D
3	ABD	C
4	EBC	D
5	BCE	D
6	ECF	D
7	CEF	D
8	ABD	C

The prism ABCDEF has been divided into three pyramids equal to one another which have triangular bases [Euclid's *Elements* XII proposition 7].

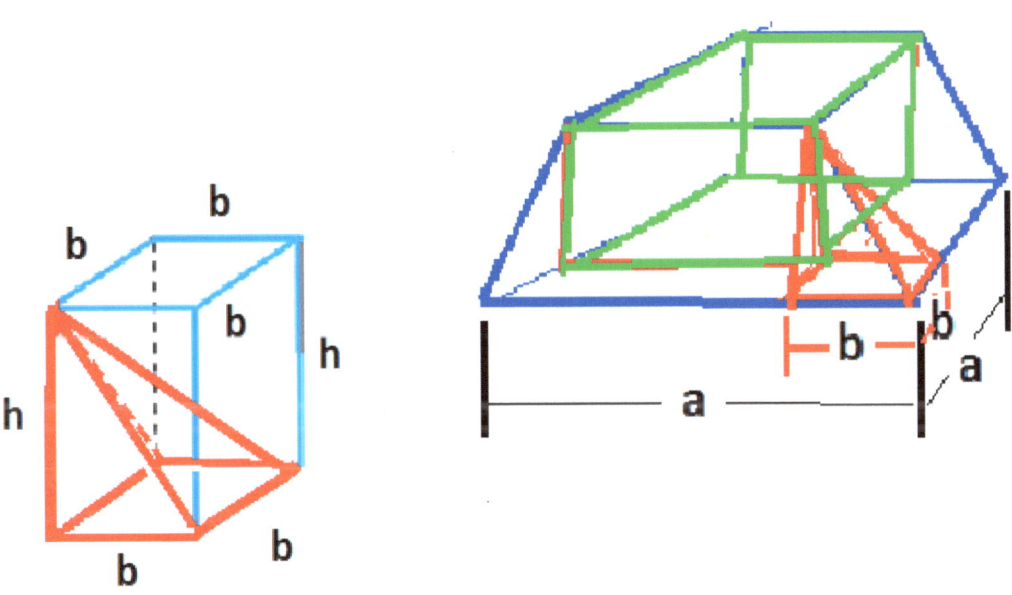

Figure 6

T. Margulies, *Pyramid Geometry Design*

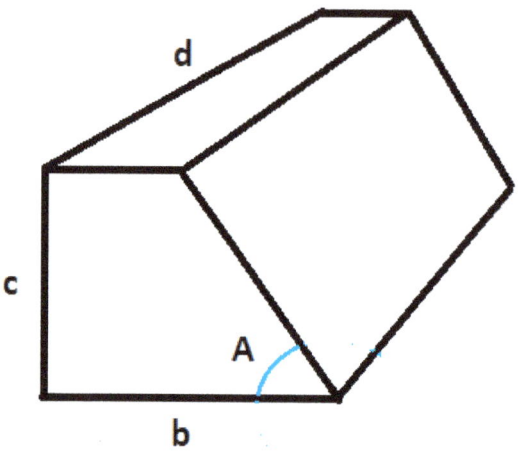

Figure 7

This geometry along with algebra derives a formula for the truncated pyramid. Summing the rectangular solid with two side wedges and a Figure 2 corner pyramid [oblique] with a rectangular base and the apex over a vertice of the base yields.

$$V = b^2h + 2\left(\frac{a-b}{2}\right)bh + \frac{(a-b)^2h}{3}$$

$$= b^2h + abh - b^2h + \frac{(a^2 - 2ab + b^2)h}{3}$$

$$= \frac{3abh}{3} + \frac{(a^2 - 2ab + b^2)h}{3} = \frac{(a^2 + ab + b^2)h}{3}$$

This derivation corresponds to that in the Moscow Papyrus, referred to as the Golenishev Papyrus, Problem 14 with numbers for an example calculation.

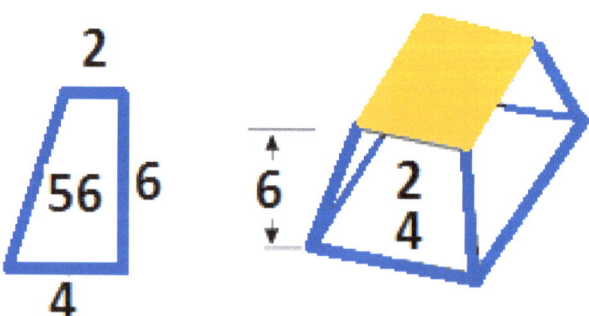

Figure 8

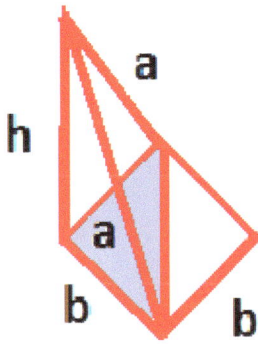

Figure 9

The geometric progression known to Egyptians may be derived simply by summing n terms and subtracting the common ratio r multiplying the sum S_n,

$$S_n = b + br + br^2 + \cdots br^{n-1}$$

$$S_n - rS_n = b - br^n$$

$$S_n(1 - r) = b - br^n$$

$$S_n = \frac{b(1 - r^n)}{1 - r}$$

Problem seventy-nine of the Rhind Papyrus presents a geometric progression sum solution. There are seven houses, in each house there are seven cats, each cat kills seven mice, each mouse has eaten seven grains of barley, each grain would have produced seven hekat. What is the sum of all the enumerated things.

$$7 + 7^2 + 7^3 + 7^4 + 7^5 = 7(1 + 7 + 7^2 + 7^3 + 7^4)$$

Table 2

houses	7
cats	7x7 = 49
mice	7x7x7 = 343
Barley (7 grains)	7^4 = 2401
heckets	7^5 =16807
Total sum	19607

The common ratio $r = 7$ and $n = 5$. Today's arithmetic-algebra derives,

$$S_{n=5} = \frac{7(1 - 7^5)}{1 - 7} = 19,607$$

T. Margulies, *Pyramid Geometry Design*

The arithmetic progression sum is solved in problem sixty-four of the Rhind Papyrus.

Divide 10 hekats of barley among 10 men so that the common difference is 1/8 of a hekat of barley.

Egyptian Solution:

Average the value 10/10 = 1. The total number of differences is then $10 - 1 = 9$.

Find half the common difference, $\frac{1}{2} \cdot \frac{1}{8} = \frac{1}{16}$.

Multiply 9 by $\frac{1}{16}$: $\frac{1}{16} + \frac{8}{16} = \frac{1}{2} + \frac{1}{16}$.

Add this to the average value to get the largest share $1 + \frac{1}{2} + \frac{1}{16}$.

Subtract the common difference, $\frac{1}{8}$, nine times, $9 \cdot \frac{1}{8} = 1 + \frac{1}{8}$, to get the lowest share.

That is, $\left(1 + \frac{1}{2} + \frac{1}{16}\right) - \left(1 + \frac{1}{8}\right) = \frac{1}{4} + \frac{1}{8} + \frac{1}{16} = \left[\frac{7}{16}\right]$. The partial sums in today's notation would be given by,

$$\left\{\frac{7}{16}, \frac{1}{2}, \frac{9}{16}, \frac{5}{8}, \frac{11}{16}, \frac{3}{4}, \frac{13}{16}, \frac{7}{8}, \frac{15}{16}, 1\right\}$$

The total S_n is 10 hekats of barley. $\frac{S_n=10}{n=10}$ is one. This solution statement shows the sum S of the first n terms of an arithmetic sequence starting with s, and common difference d,

$$[s, s + d, s + 2d, \dots s + (n-1)d]$$

Here, d $=\frac{1}{8}$, n=10, and their formulation is $\frac{S_n}{n} = s + \frac{1}{2}d(n\text{-}1)$ or substituting and rewriting,

$$\frac{S_n=10}{n=10} = 1 = \frac{7}{16} + \frac{1}{2}\frac{1}{8}(10\text{-}1) = \frac{7}{16} + \frac{9}{16} = 1$$

Alternatively, the familiar formula is given as, $S_n = sn + \frac{1}{2}dn(n\text{-}1)$

Egyptian fractions were written with a numerator of one with the exception of $\frac{2}{3}$ as found in the Rhind Papyus as shown in the first problems of dividing ten loaves among one through ten people.

A pyramid viewed mathematically by a series summation of layers is given by,

T. Margulies, *Pyramid Geometry Design*

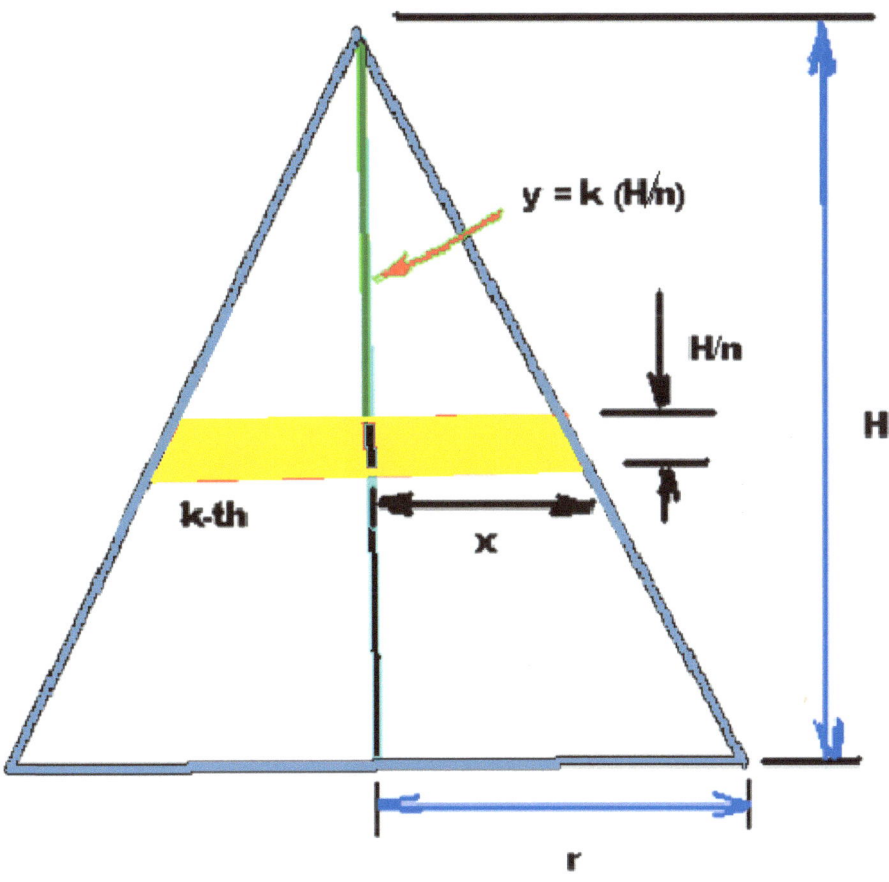

Figure 10

By similar triangles $\quad \frac{x}{y} = \frac{r}{H}$, so that $x = \frac{r}{H} y = \frac{r}{H}\left(k\frac{H}{n}\right) = \frac{k \cdot r}{n}$

$$(2r)^2 \frac{H}{n} = \frac{r^2 k^2}{n^2} \frac{H}{n} = \frac{r^2 k^2 H}{n^3}$$

$$\frac{r^2 H}{n^3}(1^2 + 2^2 + 3^2 + (n-1)^2) \quad =$$

$$\frac{r^2 H}{6n^3}(n-1)\cdot(n)\cdot(2n-1) = \frac{r^2 H(2n^3 - 3n^2 + n)}{6n^3} = \frac{r^2 H}{6}\left(\frac{2n^3}{n^3} - \frac{3n^2}{n^3} + \frac{n}{n^3}\right)$$

The limit as n approaches infinity $\qquad \frac{r^2 H}{6}\left(2 - \frac{3}{n} + \frac{1}{n^2}\right) = \frac{2r^2 H}{6} = \frac{r^2 H}{3}$

The following summations which can be substantiated by proofs of induction were used in the derivation. $\sum_{i=1}^{n} i^2 = n\cdot(n+1)\cdot(2n+1)/6, \ \sum_{i=1}^{n-1} i^2 = (n-1)\cdot(n)\cdot(2n-1)/6$

Concurrent in their design is the use of the Golden mean, cut, section, or ratio. The Greek letter Phi for the first letters of the sculptor Phidaeus credited with the Parthenon architecture in Athens is the popularly adopted name. Previous Greek letter tau was used for the Golden mean. Later the Golden portion is presented to emphasize a common name for a human vision's yellow light portion of the perceived spectrum.

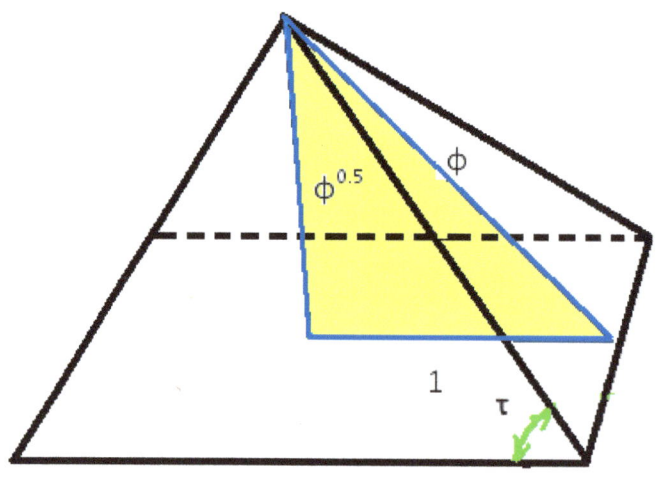

Figure 11

Using Pythagorus' theorem on right triangles,

$$\phi^2 = \phi + 1 \quad \text{or } \phi, \text{ Phi, is a solution of the quadratic equation} \quad \phi^2 - \phi - 1 = 0$$

Table 3

Phi	1.61803399
(Phi)^0.5	1.27201965
1/Phi	0.61803399

The musical scale called the minor Sixth, A flat, is given by $5/8 = 0.625 \approx \frac{1}{\phi}$.

The ratio of base to height is approximately equal to Phi as shown in the following table for the great Gizah pyramids. The ratio of volumes of the first and third is about two or an octave. The measures of royal cubit equals seven palms which equal twenty-eight fingers.

$$Tangent(Angle) = \frac{\phi^{0.5}}{1} \ , \ Cotangent(Angle) = \frac{1}{\phi^{0.5}} \text{ where } \phi^{0.5} = \sqrt{\phi}.$$

Table 4

T. Margulies, *Pyramid Geometry Design*

	Phi	Angle	Phi^0.5	Angle	1/Phi^0.5	Angle
	Radians	Degrees	Radians	Degrees	Radians	Degrees
Tangent	1.017222	58.28257	0.904557	51.82734	0.666239	38.17274
Cotangent	0.98307	56.32579	1.105514	63.34132	1.500962	85.99885

The Egyptian Seked which corresponds to the horizontal change to the vertical is similar to a cotangent function calculation for angle by ratio of opposite and adjacent side-lengths.

Table 5

Name	Angle	Date [BCE]	Date [BCE]	Base	Height	Vol.	Vol. Ratio	Base/Ht	7xBase/(2xHt)
Khufu Akhet	51° 51' 14"	2170	2170	9168	5820	17785920		1.575	5.513
Khafre Wer	52° 20' 00"	2130	2130	8493	5451	16476420	2.155148	1.558	5.453
Menkaure Netery	51° 00' 00"	2100	2100	4254	2616	8252760	1.996474	1.626	5.692

Table 6

Name	Brightness magnitude
Khufu	1.74
Khafre	1.69
Menkaure	2.25

The square root idea is believed to be known to Egyptians. It derives from the idea that a square whose area is A has side-length or root $\sqrt{A}$.

An early idea for a square in deriving a modified π (Pi) as the square perimeter dived by a side-length appears as an alternative to that of the circle which is a circumference divided by diameter for any size circle. Knowledge of the $\frac{22}{7} = 3.14286$ approximation for Pi was considered unlikely by N. Wheeler.

Archimedes in *Measurement of a Circle* derived a formula for the area of a circle. The circle geometry estimate for π (Pi) is approximately 3.14159 . The Rhind Papyrus estimates Pi as 3.16149 . All estimates are approximations as a decimal representation sine Pi is an Irrational real number. Only

T. Margulies, *Pyramid Geometry Design*

Rational numbers have a finite number of digits to the right of the decimal point or repeating unit of digits.

Eratoshenes (275-194 BCE) is famous for the first estimate of the Earth's circumference to approximately within two percent error and contributed the leap year concept and finding of prime numbers. Syene is situated at the Tropic of Cancer, north of the Equator, where the Sun reaches its most northerly point at noon during the summer, the summer Solstice at June 21-st at noon. The Sun when overhead in Syene (Aswan) it casts a shadow angle of approximately 7.2 degrees in Alexandria, 5000 Stadia distance units away. Angle was calculated in Seked units as the number of palms per one vertical cubit rise. The ratio and proportion problem for a 360 degree circle as the polar diameter of Earth becomes,

$$\frac{7.2}{360} = \frac{5000}{C} \quad or \quad C = \frac{360}{7.2} \cdot 5000 = 250{,}000 \text{ Stadia}$$

The Earth's polar circumference estimate is $250{,}000$ Stadia $24{,}461\ mi\ \ or\ \ 39{,}186\ km.$

Numerous astronomical observations are recorded and reviewed by Piazzi Smyth. For example, the sighting of the Constellation Draco, or Draconis the dragon, aligns with the subterranean chamber entrance passage with the winter solstice 2170 BC when crossing below the pole [PlateVI]. The winter solstice is the day on which the Sun's shadow is the longest. The Orion Constellation aligns as well. Early Zodiac designs consisted of probably six constellation signs. Sirius, α-Canis Majorus or *dog* Star, which is both bright and close to Earth had noticeable helical risings minutes before dawn to mark the flooding of the Nile and the Egyptian calendar.

The building of pyramids grew as depicted in Figure 9 listing 192 [N. Wheeler]. Many measurements on slopes were missing or owners questionable; however the large listing indicates a passion to achieve their purpose.

The Great Giza Pyramids rely heavily on its use in design. Analyzing forty shown in Figure 10 as a distribution of angles indicate a non-adoption of the Golden Phi design as the optimum in other design motivations. The nvestigations reported by Noel Wheeler shows the frequency of pyramid building timeline and a pyramid slope size list. A statistical frequency and cumulative distribution along with some summary statistics are shown in Figure and Table . An alternate Egyptologist, Charles Piazzi Smyth has given extensive data and measurement analysis. The slope, height, and base data are analyzed as displayed in Figures 12 through 16 and accompanying tables.

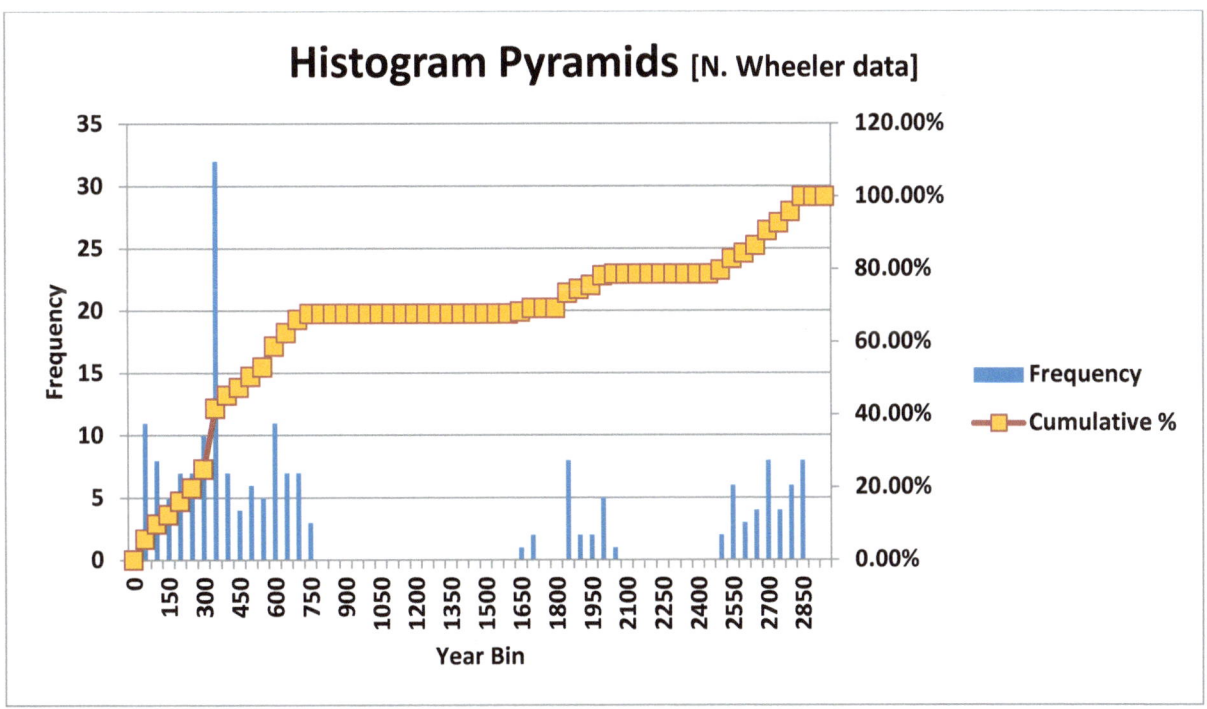

Figure 12

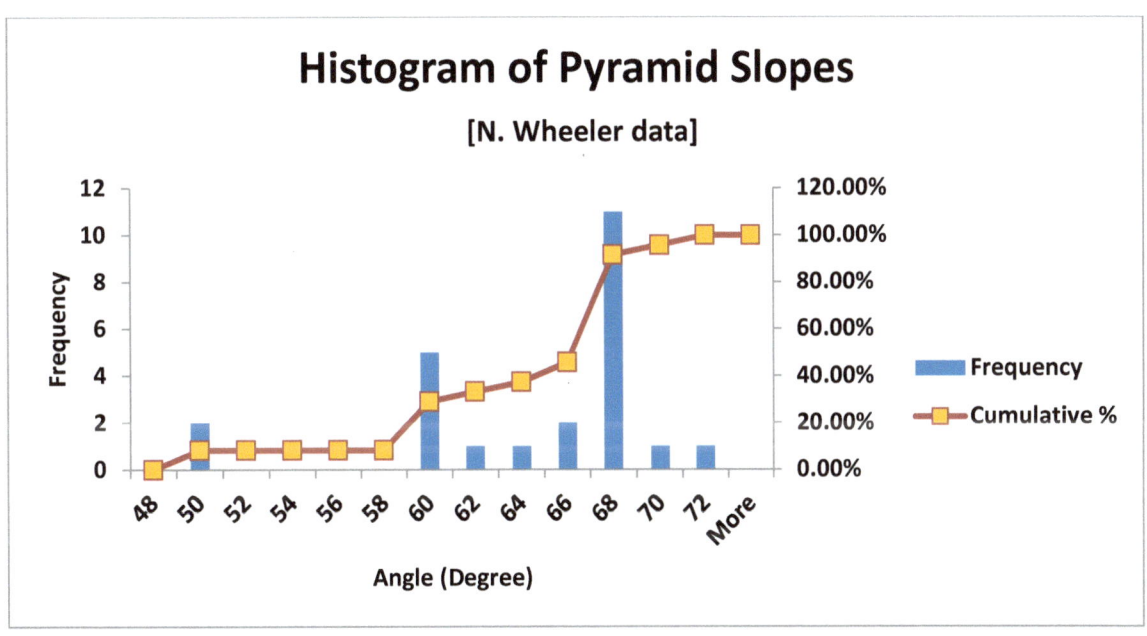

Figure 13

Table 7

Average (n=240	64.302
Median	67.25
Maximum	72
Minimum	48.75
Std. Dev.	5.833

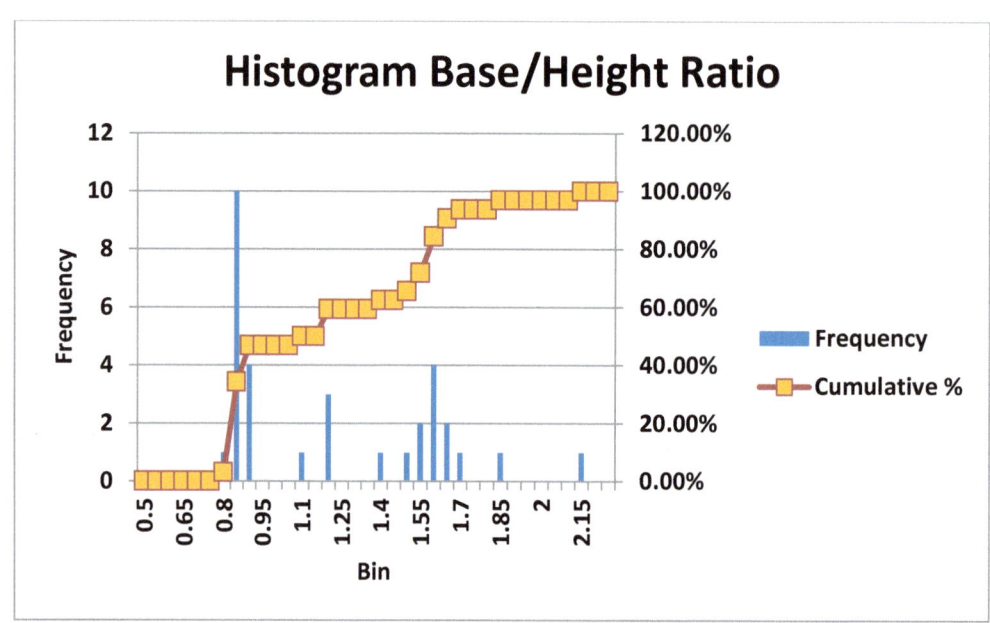

Figure 14

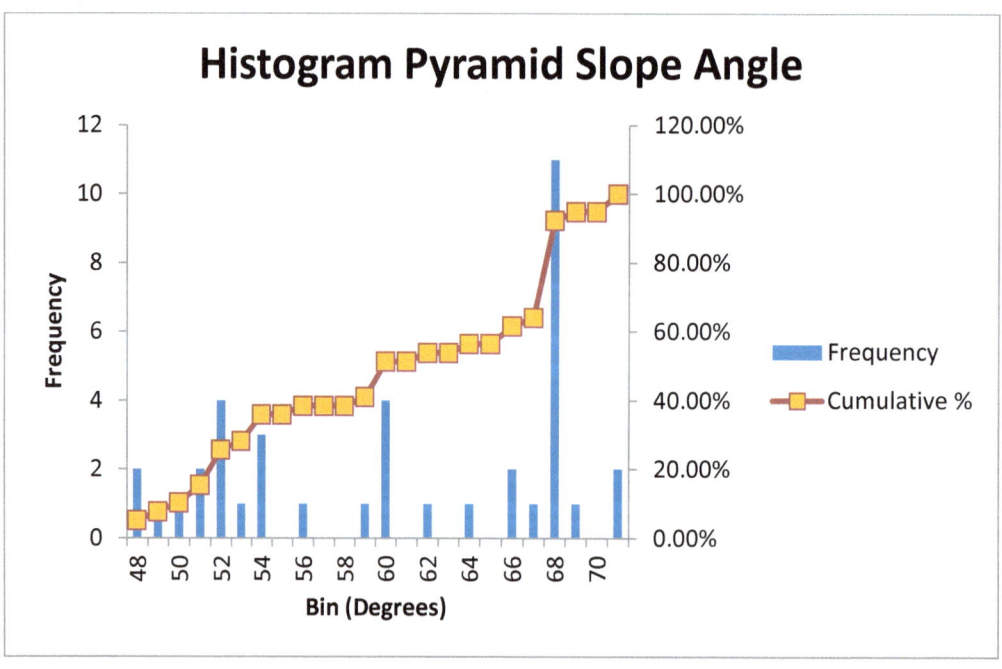

Figure 15

T. Margulies, *Pyramid Geometry Design*

Table 8

60.458	Ave (n=39)
60	Median
69	Max
48.75	Min
7.237697	Std. dev.

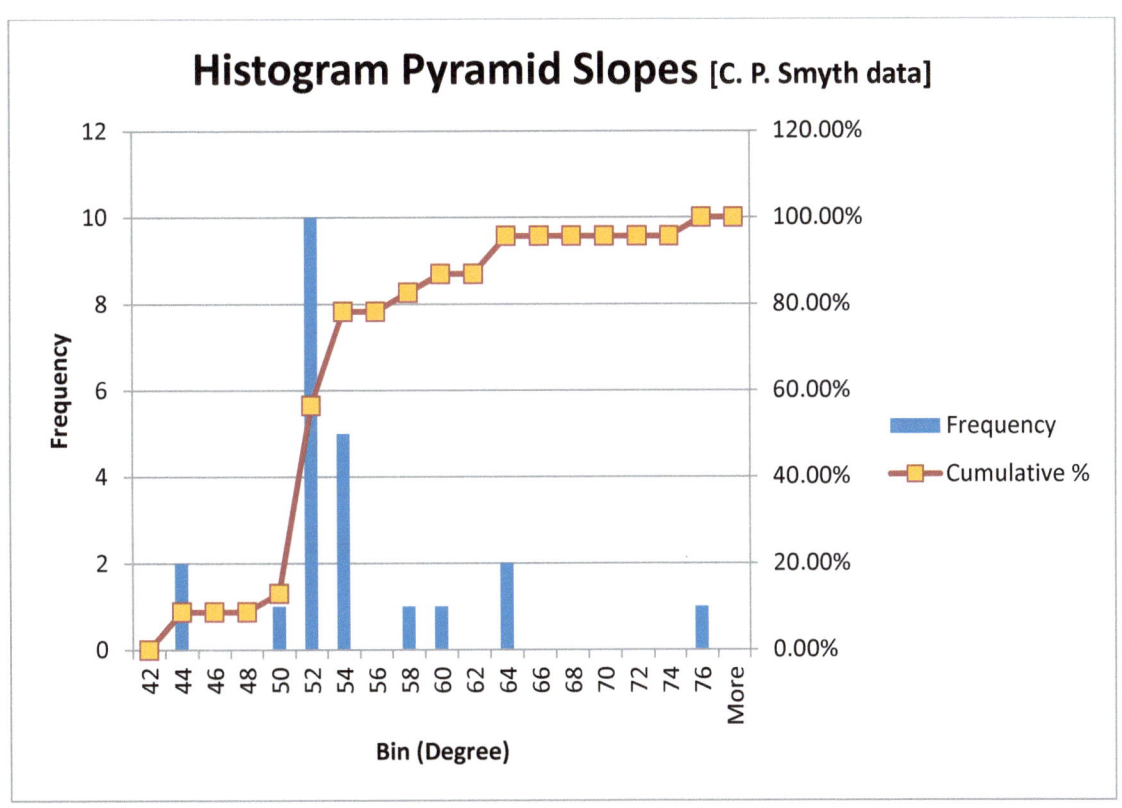

Figure 16

Table 9

53.534	Average (n=30)
52	Median
74.167	Max
42.991	Min
6.592	Std. dev.

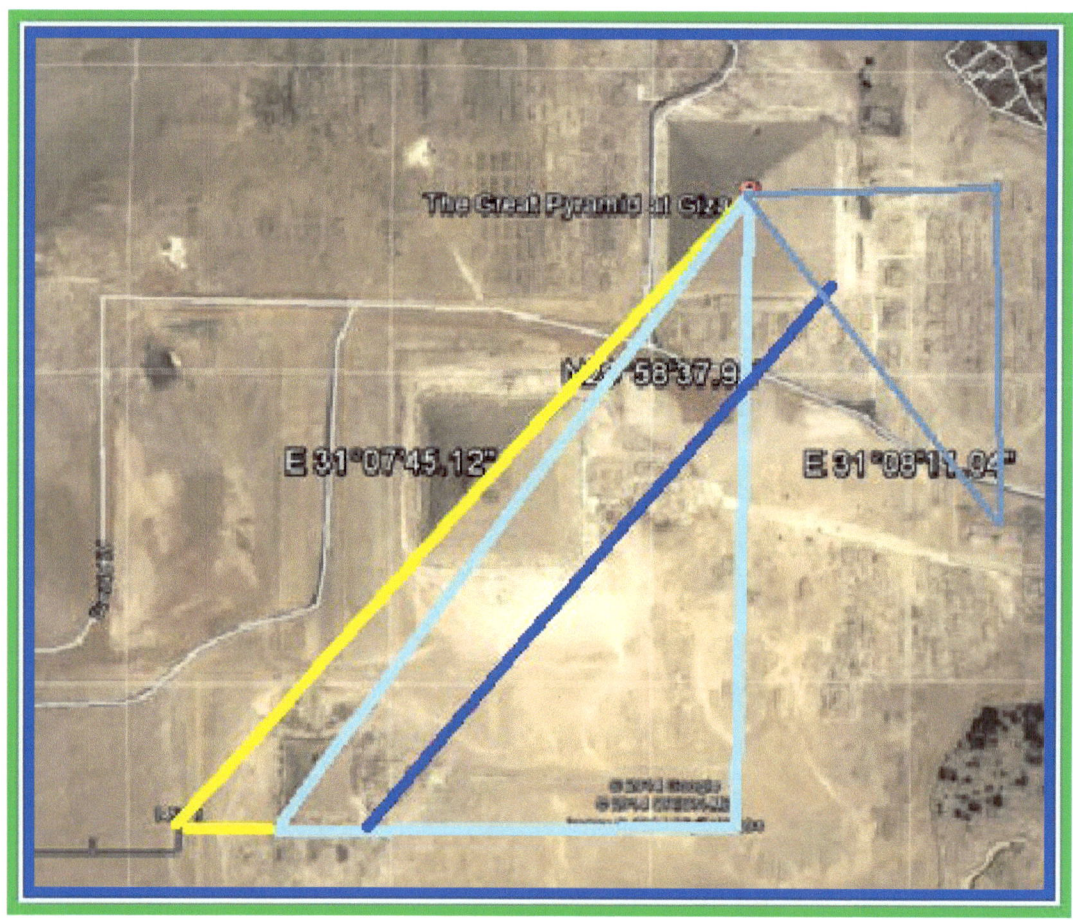

Figure 17

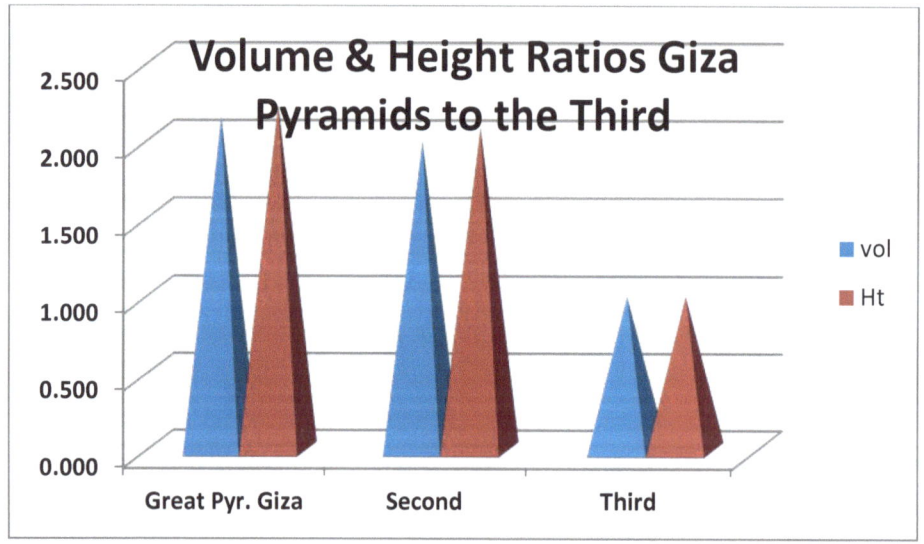

Figure 18

T. Margulies, *Pyramid Geometry Design*

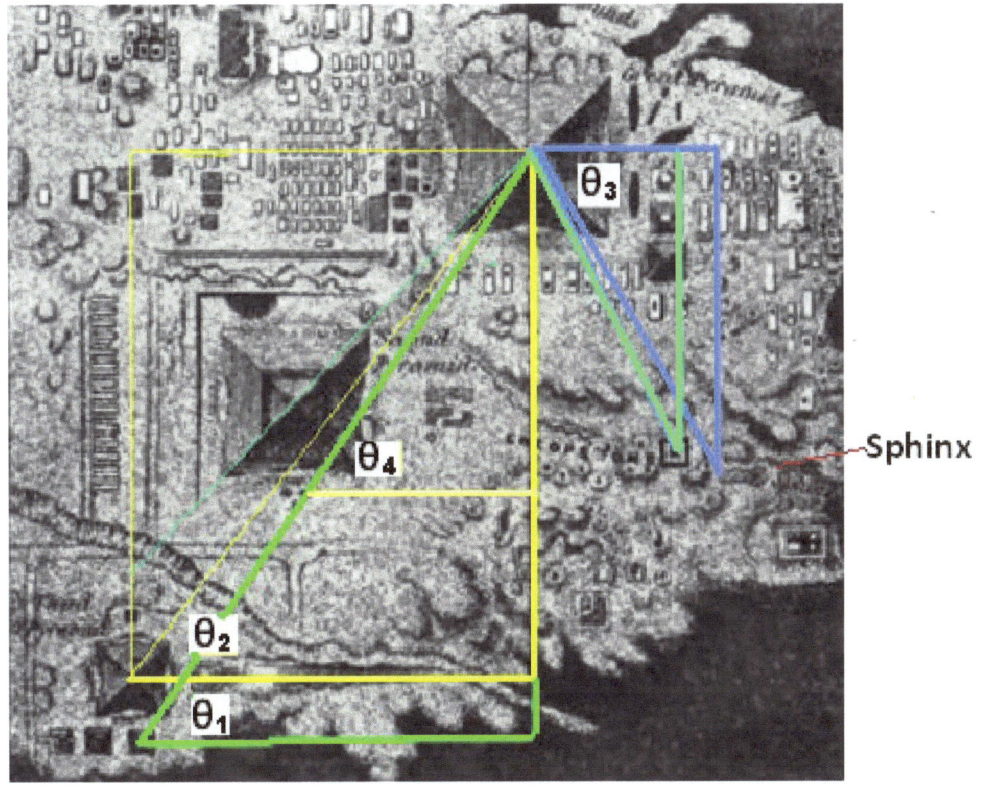

Figure 19

Plate III: Map of The Pyramids of Jeezeh [Piazzi Smyth, *The Great Pyramid*, A. Ritchie & Son, Edin.]

A simple comparison of the Khufu Pyramid with that of the temple of the Sun is tabulated below.

Table 10

Measurement/Calculation	Khufu	Sun
Height	481.3949	233.5
Side(base)	755.79	733.2
Height/base	0.63694267	0.318467
Slope Angle	51.827	32.494
Ht/(0.5*Side)	1.27388534	0.636934
Base Perimeter/Height	6.28000006	12.56017
Ratio of base Areas	1.06256956	0.941115
Ratio of base Perimeters	1.03081015	0.970111
Slant Side Angles τ	58.3057932	49.8543
Perimeter	3023.16	2932.8
Diagonal d=Sqrt(2)*side	1068.84847	1036.901
2*side^2	1142437.05	1075164
d^2	1142437.05	1075164

T. Margulies, *Pyramid Geometry Design*

The pyramids Khufu and the Sun have volume ratios compared to volumes formed by a square base and height as a rectangular solid and a cube of their side-lengths that are approximately one-half Pi and Pi, respectively.

Solomon's Temple is a 2/1 ratio (length/height) or musical octave. 60 cubits length x20 cubits long x 30 cubits high = 36000 cubic cubits.

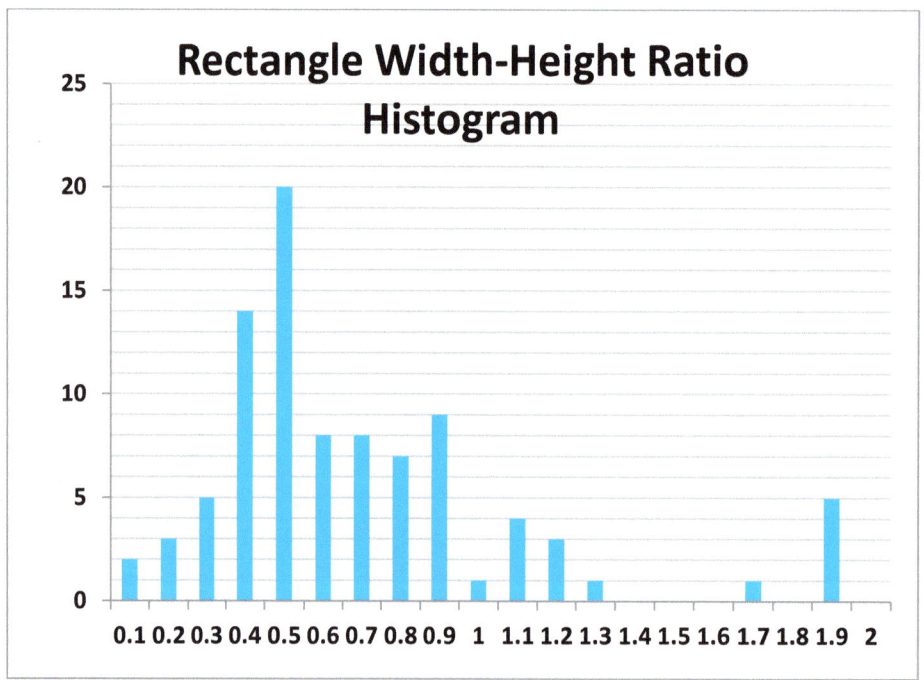

Figure 20

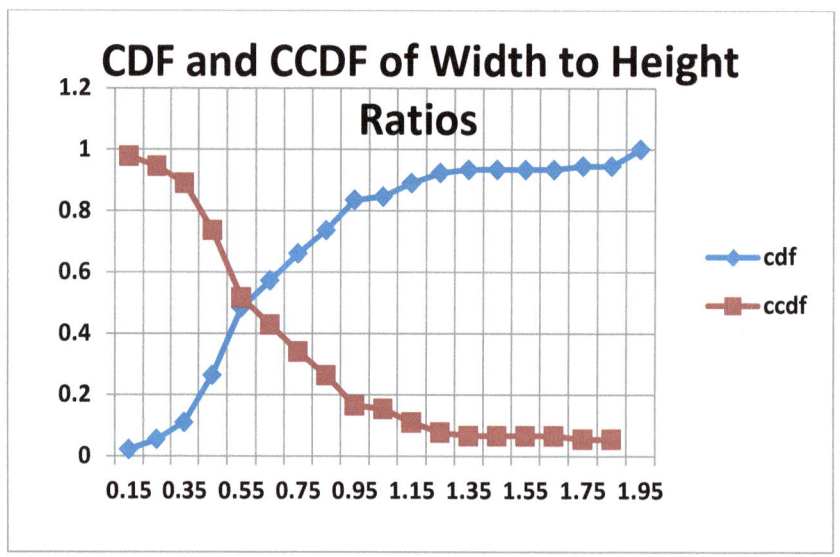

Figure 21

T. Margulies, *Pyramid Geometry Design*

On Luminosity and Pyramid Arrangement at Giza

These Stars in Orion's in Egyptian were named Pharoah-God as represented as three. In Greece the constellation is known as that of the Greek hero and hunter Orion who was killed by a Scorpion. In ancient Egypt, Orion's belt was known as the symbol of the God-pharaoh, Osiris. Apparently, the name and ideas shifted from an agricultural Corn God to God of creation, anticipation, and resurrection in a Earth-Sun-Moon universe.

Figure 22

Orion's belt consists of three stars, Mintaka, Alnilam, and Alnitak. Mintaka means belt, Alnilam means "a belt of pearls", and Alnitak means the girdle in Arabic. Mintaka is a super-giant star and is white blue and 10,000 times brighter than our sun. Alnilam which is the center star of Orion's belt is also a white blue super giant is the brightest of Orion's belt stars at 18,000 times brighter than our Sun. The third star, Alnitak is the dimmest of the stars. The Sun is white in color and temperature; however, appears as yellowish from atmospheric scattering of the smaller wavelength blue light. The Rayleigh and Mie scattering phenomena alters directions of the incoming rays. In a red, green, blue primary system of colors, yellow is *minus blue*. The great Giza pyramid is was white becoming cream with age.

T. Margulies, *Pyramid Geometry Design*

Using the three Giza Pyramid volumes and height estimates linear regressions were made with an $R^2 \sim 0.96$. Two points determine a perfectly staright line and the analysis uses three. The display and human visual perception is logarithmic.

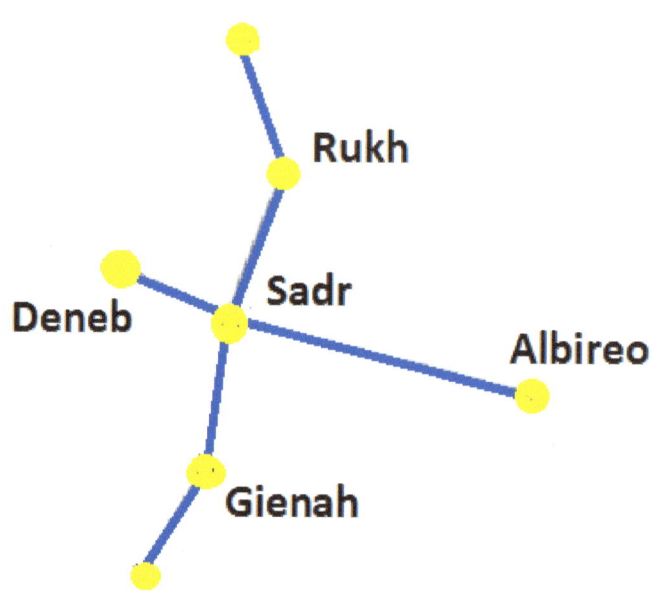

Figure 23

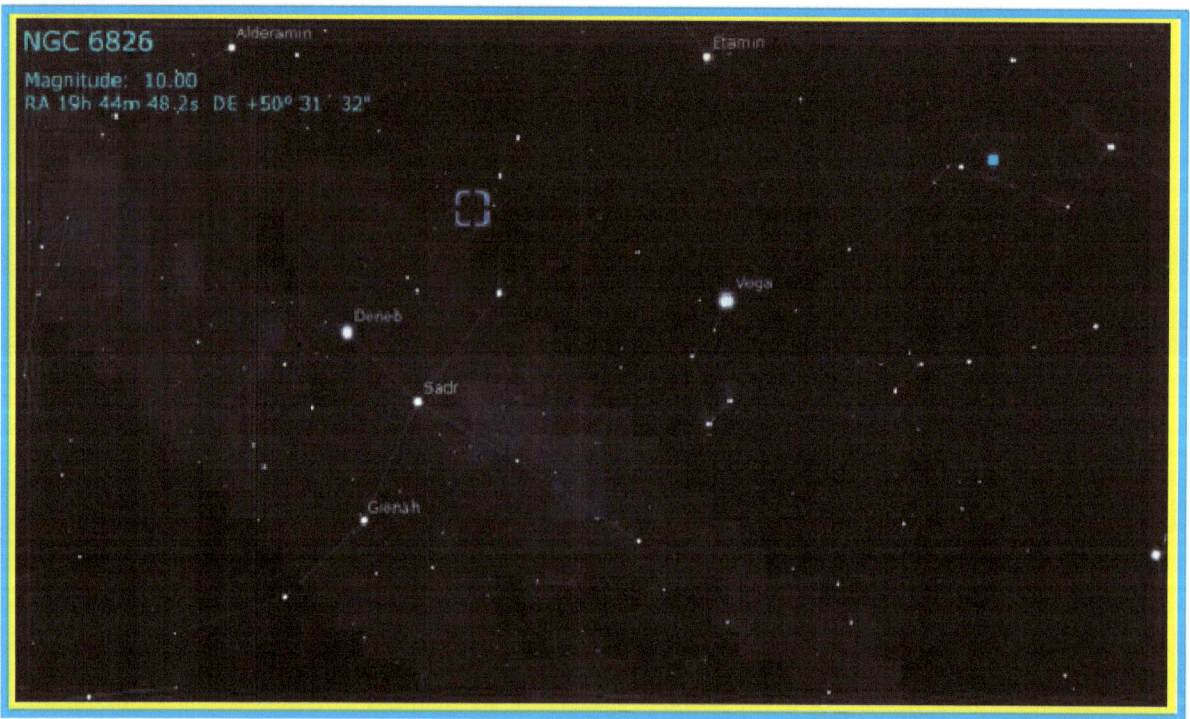

Figure 24

T. Margulies, *Pyramid Geometry Design*

Additionally, the Cygnus or Swan constellation has a bright star wing arrangement as another star candidate for investigation. The apparent magnitudes of the stars, Rukh, Sadr, and Gienah were also used in a linear regression with $R^2 \sim 0.90$. The slopes for magnitude versus pyramid volume are approximately the same. Rukh is a triple sar system comprised of a bluish-white giant, a yellow-white, and orange sequence. Sadr is bright yellow with a radius of 150 times larger than the Sun. Gienah is an orange giant about eleven times larger than the Sun with twice the mass.

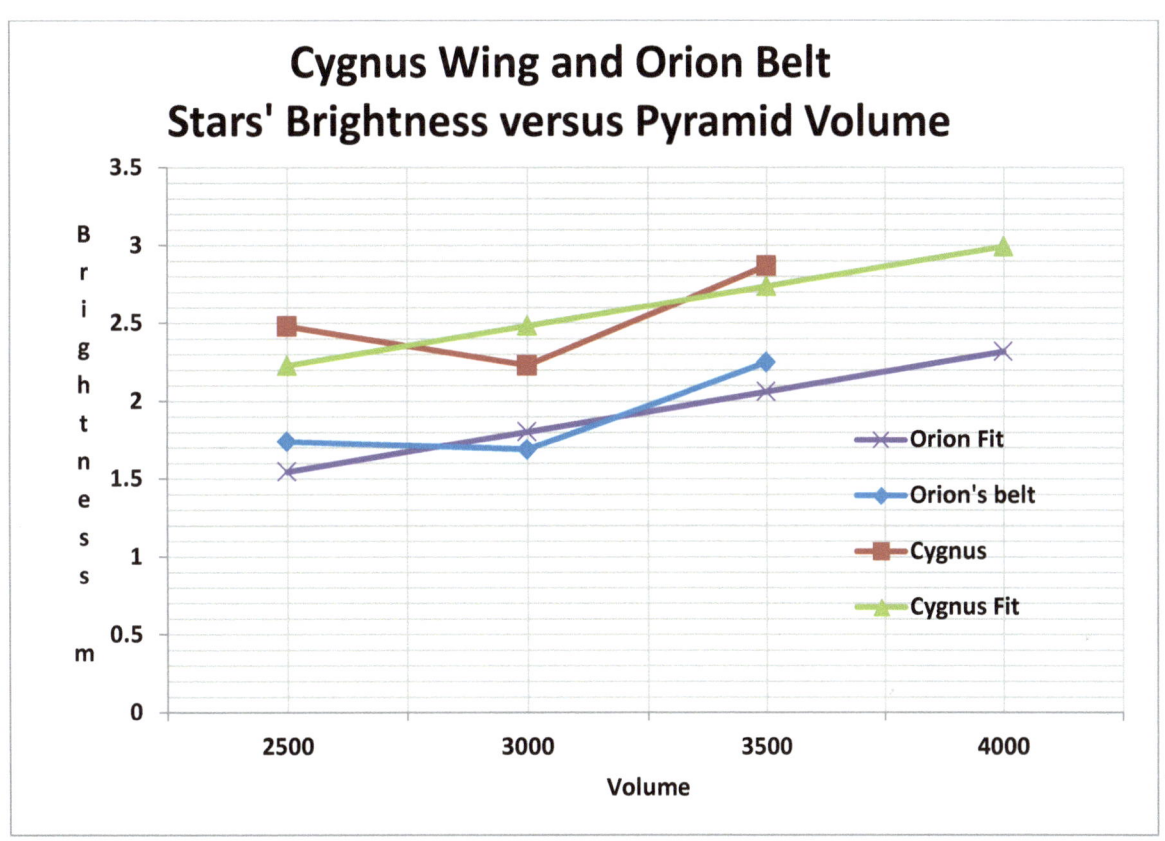

Figure 25

Pyramid data points used are listed.

Table 11

name	Latitude	Angle	Date [BC]	Base	Height	Vol.
Khufu	29° 59'	51° 51' 14"	2170	9168	5820	17785.92
Khafre	29° 59'	52° 20' 00"	2130	8493	5451	16476.42
Menkaure	29° 58'	51° 00' 00"	2100	4254	2616	8252.76

Hipparchus was a Greek astronomer who subjectively ranked star brightness on a scale of one through six around the date of 150 BCE. One was assigned to the brightest and six to the dimmest. An increase in five magnitude was probably associated with a decrease of one-hundred.

Table 12

Name	Orion Belt Stars Brightness m	Cygnus Stars Brightness m
Khufu	1.74	2.48
Khafre	1.69	2.23
Menkaure	2.25	2.87

The starlight magnitudes, absolute [M} and relative [m] are calculated in recent astronomy by a flux that accounts for the number of photons passing an area and time. Formulas for absolute and relative star brightness magnitudes are given by M, and m respectively. Astronomical range or distance is denoted by d and time in parsec by pc.

$$M = m - 5 \cdot \log\left(\frac{d}{10\ pc}\right)$$

Energy flux, F, at a given luminosity, angular size distance, and motion Doppler red-shift is a function of frequency. For a uniform spherical bright source of radius r with the values integrated over all frequencies the energy flux varies as $(1 + z)^{-4}$; that is, $F = \int_0^1 2\pi I \cdot \cos(\theta)\, d(\cos(\theta)) = \pi I$.

Furthermore, The luminosity is the multiplicative product of the surface area of the sphere and F.

$$L = (4\pi r^2)\pi I = 4\pi^2 r^2 I$$

The universe expands making the observed surface brightness given by $I_0 = \frac{I}{(1+z)^4}$ and the angular radius of the source at the observer is $\theta = \frac{r(1=z)}{a_0 \cdot r(z)}$. Then the observed energy flux is the product of the surface brightness and solid angle, $F = \pi\theta^2 I_0$ and $F_0 = \frac{L}{4\pi(a_0 \cdot r)^2(1+z)^2}$. As the source radius increases which is in the denominator the energy flux decreases.

Hipparchus during the second century made naked-eye Star observations and devised a six scales of

T. Margulies, *Pyramid Geometry Design*

brightness categorization. Later photometric studies of the average range in brightness between scales one and six is one-hundred. Norman R. Pogson in 1856 chose the difference between two magnitudes should correspond then to $(100)^{\frac{2}{5}} \approx 2.512$. A logarithmic formulation for the brightness of two stars is given as, $\frac{b_2}{b_1} = 10^{0.4(m_1-m_2)} = 2.512^{(m_1-m_2)}$. It is noted that the physical scale would be similar to the acoustic bel or decibel if one or ten were chosen rather than the four-tenths.

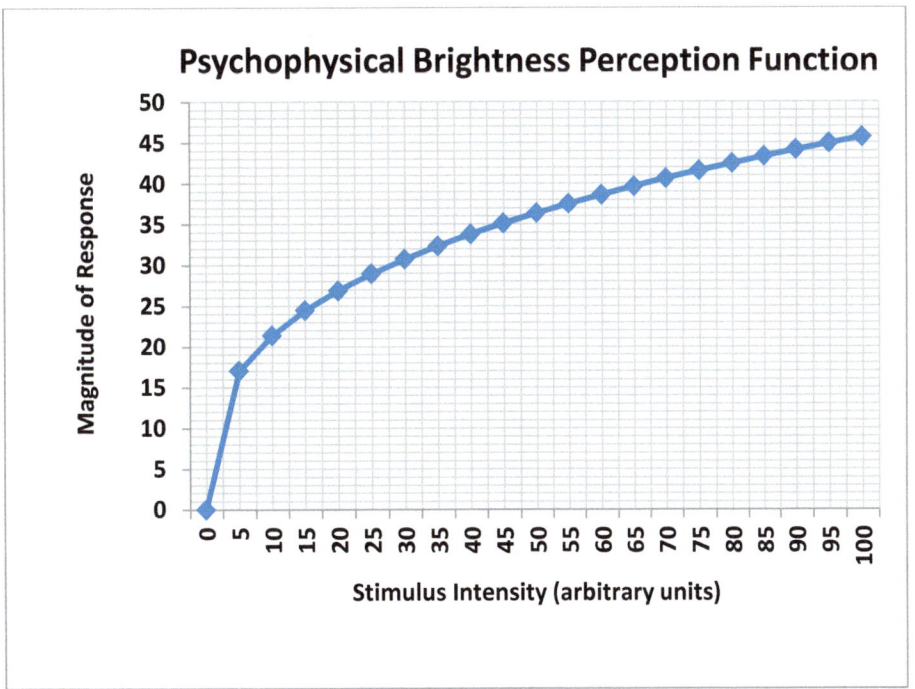

Figure 26

The plotted power function for physical stimulus S is given by, response

$$R = 10 \cdot S^{0.33} \text{ or equivalently } S = \left(\frac{R}{10}\right)^3 = 0.001 \cdot R^3 \text{ and } \log(S) = \cdot 3\log(R) - 3.$$

The understanding of the perception of visual sensory response to light intensity has beginnings in psychophysical research by Gustav Fechner (1860) and Ernst Weber (1795-1878). The latter, for example, expressed an understanding of just noticeable differences " in observing the disparity between things that are compared, we perceive not the difference, but the ratio of the difference to the magnitude of the things compared". The Fechner law based on logarithms is just mathematically the inverse function of the exponential power function. That is, the change in intensity of stimulus in ratio to a constant magnitude is considered constant. This is easily represented by a power function or its inverse logarithmic function. Data by S. S. Stevens (1957) for a five degree lighted spot in the dark

T. Margulies, *Pyramid Geometry Design*

fits an intensity power function with approximately a one-third exponent. The Sun or moon is approximately thirty minutes (0.5 degrees).

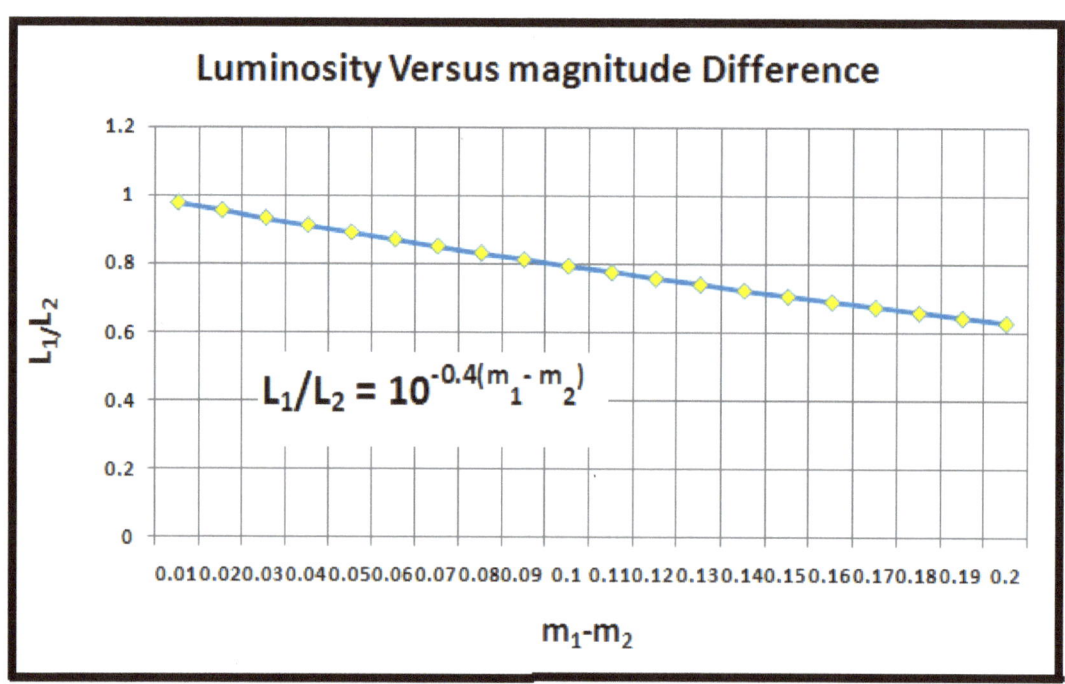

Figure 27

A lens power, P, is reciprocal to the focal length, $P = \frac{1}{f}$ so that a shorter focal length corresponds to a stronger lens to redirect the light ray-paths.. For focal lengths given in units of meters [m], the power has units of reciprocal m or diopters.

Table 13: Alternative Light Measurement Units

Generic		
Amount of Light / Time	Photons/sec	Lumens-sec [lm-s]
Amount of Light / Time/solid angle	Photons/sec/sr	Lumen/sr [lm/sr]= candelas [cd]
Amount of Light / Time/ Area	Photons/sec/m^2	Lm/ m^2 =lux [lx]
Amount of Light / Time/solid angle/ Area	Photons/sec/sr/m^2	Lm/sr/ m^2 = cd/ m^2

troland [tr]: a unit for retinal illumination viewing surface of lamination 1 cd/ m^2 through the pupil 1 mm^2; sr = steradian.

Psychology of vision research on reflectance of a light spot with a constant background light level has been studied for subjective choice of the lighter. The resulting probability distribution of the results in a cumulative distribution form are presented in the Figure 28 [Allred and Brainard, 2008]. A psychophysical receiver operator characteristic (ROC) summary of type I and II statistical errors are helpful in data analysis when the background has a variable noise compared to the light signal discussed by Green and Swets.

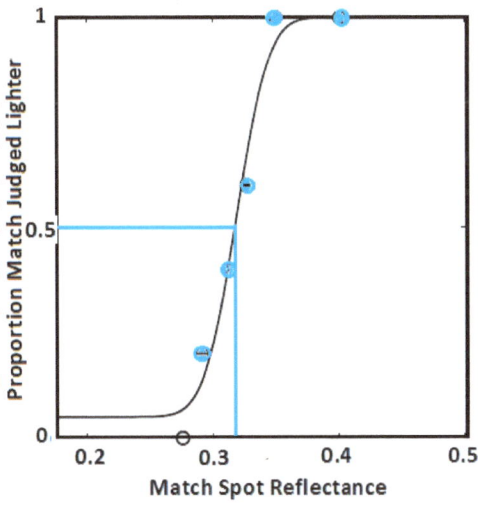

Figure 28

Global Monument Perspective

The pyramids and monuments when plotted on a globe show an approximate great circle. Here an equation for a plane is derived and used to represent an equation for fitting the data of coordinate positions in a three dimensional space. Let $\underline{x_0}$ be the position vector of the given point in the plane, and let $\underline{n}$ be the given normal vector to the plane.

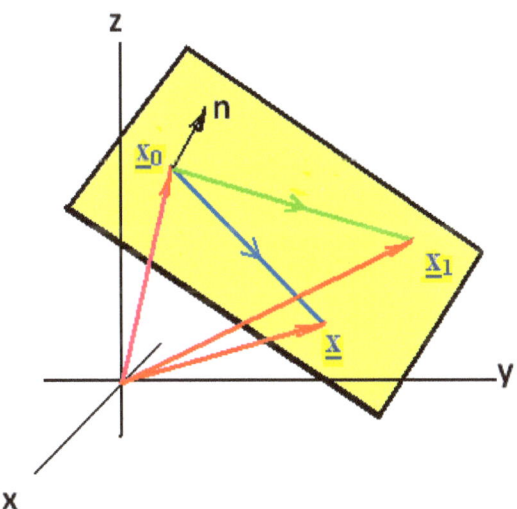

Figure 29

T. Margulies, *Pyramid Geometry Design*

Let $\underline{x}$, $\underline{x}_0$, $\underline{x}_1$ denote vectors points in the plane. Writing coordinates for $\underline{x}_0 = (x_0, y_0, z_0)$,

$\underline{x}_1 = (x_1, y_1, z_1)$ the equation of the plane may be expressed by,

$$\alpha x + \beta y + \gamma z = \delta \quad \text{where}$$

$\delta = -\alpha x_0 - \beta y_0 - \gamma z_0$ with the axis intercepts given by $\quad x = -\dfrac{\delta}{\alpha},\ y = -\dfrac{\delta}{\beta},\ z = -\dfrac{\delta}{\gamma}.$

Data regression fits obtain,

$$z = -0.07337931208 - 0.9373181662\,x + 1.937336682\,y \quad \text{or}$$

$$0.9373181662\,x - 1.937336682\,y + z + 0.07337931208 = 0$$

Figure 26

Cartesian coordinates may be retrieved from the spherical coordinates (*radius r, inclination ϑ, azimuth φ*), where $r \in [0, \infty)$, $\varphi \in [0, 2\pi]$, $\vartheta \in [0, \pi]$, $x = r\ sin\theta\ cos\varphi$, $y = r\ sin\theta\ sin\varphi$, $z = r\ s\ cos\theta$

The great circle of light markings of pyramids including several other archaeological monuments, such as Rapa Nui, or Easter Island, and Stonehenge are displayed in Figure 13.

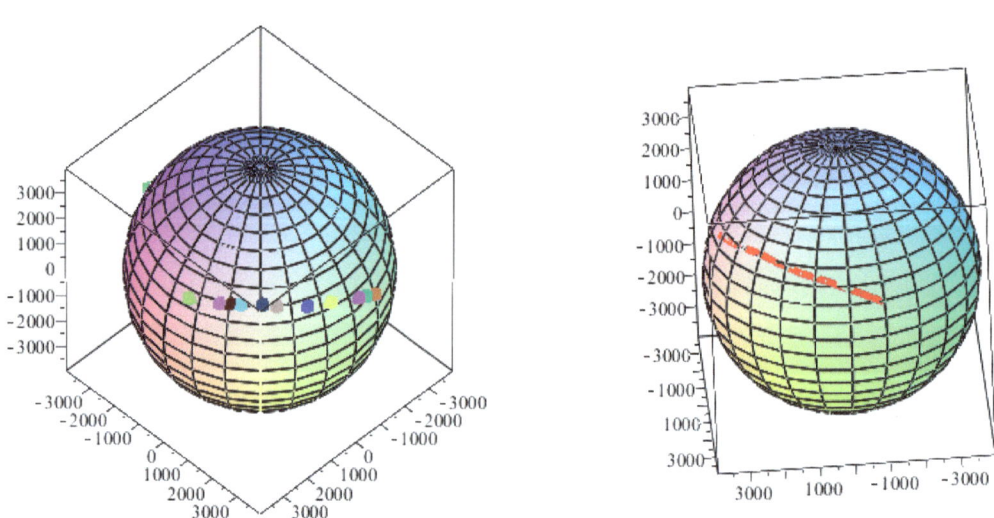

Figure 30

The closest approach of the earth to the sun is called perihelion, occurring in January, Changes in timing of perihelion is known as the precession of the equinoxes, and occurs on a period of about 22,000 years. 11,000 years ago, the perihelion occurred in July which is attributed to causing more

T. Margulies, *Pyramid Geometry Design*

severe seasonal changes. There are also 41,000 year tilt cycles and 22,000 year precession cycles. The Earth's wobble that usually accompanies the precession movements appears to have ceased relatively recently during the last several decades.

The Earth's 23.5 degree tilt is associated with the change of seasons and the day night cycle is from its spin along the vertical axis. The Earth exhibits a wobble in its orbital motion around the Sun which varies from 3 to 15 meters in generally a seven year cycle. This motion associated with a non-spherical spinning object was first studied by Isaac Newton called free nutation. A nearly ceased wobble has occurred about 2006 after the location of the Earth's spin axis made a sharp angle with implications for increased tectonic activity.

Astronomers learned that tracking the Moon to understand its cycles needed clarity of reference. New phases of the Moon or the time duration between two full Moons was approximately 29.5 days, called the synodic month. The time it takes the Moon to return to the same location relative to the stars was two days shorter in length of time, called a sidereal month. Dragons were conceived of as eating the Moon causing its change of size or phase. The Moon's lunar eclipses were first proposed by Anaxagoras of Greece around 450 B.C. *"Anaxagoras, in agreement with the mathematicians, held that the moon's obscurations (phases), month by month, were due to its following the course of the sun by which it is illuminated, and that the eclipses of the moon were caused by its falling within the shadow of the earth, which then comes beween the sun and the moon, while the eclipses of the sun where due to the interposition of the moon."* [Aetius' writings as an historian]

Also, Aristotle in about 350 B.C., observed that the curved umbral shadow of the Earth on the Moon during a lunar eclipse was a proof that the Earth was in fact spherical.

Earth has many archaeological mysteries that delight viewers' imaginations and challenge scientists for rational explanation. Several are highlighted to arouse interest, speculation, and perhaps some experimentation. Pyramids and pyramidal monuments were a global practice, found in Egypt, Peru, Central America, America, China, Japan, as well as others, France and Polynesia.

In Peru are the amazing geo-glyphs or Nazca Lines patterned as insects, animals, or creatures drawn in the desert as shown and estimated to have been created around 100 BC. The Nazca lines reside in a desert area between Palpa and Cahuachi. On the northern coast of Peru were the Moche people who are noted to have built some of the largest buildings such as the Temple of the Sun.

T. Margulies, *Pyramid Geometry Design*

Figure 31

Figure 32

T. Margulies, *Pyramid Geometry Design*

The image in the above picture shows a similar alignment of pyramids with probable astronomical motivations.

Linear Section Golden Proportion Phi

A simple derivation is articulated as the equality of two ratios in the form of line segment lengths, commonly referred to as a proportion.

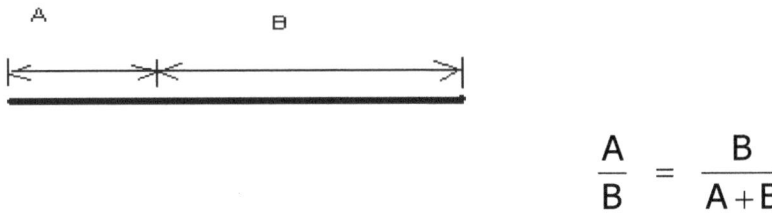

$$\frac{A}{B} = \frac{B}{A+B}$$

Figure 33

Let $x = \dfrac{B}{A}$ to write the familiar quadratic equation in the ratio of lengths B and A,

$$x^2 - x - 1 = 0$$

In general the solutions are, $x = \dfrac{1+\sqrt{5}}{2} \approx 1.61803$, $x = \dfrac{1}{2} - \dfrac{\sqrt{5}}{2} \approx -.61803$

This derivation corresponds to the interpretation of the length to width ratio of line segments which are like the side-lengths of a rectangle, the longer side is approximately 1.61803 multiplied by the shorter side length. These are named as Phi ratios. Phi ratios exhibit themselves in humans and organic structures such as the length ratios in, for example, Egyptian pyramids and Parthenon architectures, the human body, butterflies, horses, dragonflies, and some fish. The Golden ratio is a is an irrational number named *Phi* as the first Greek letter in the sculptor Phidias who is known for his masterpiece Zeus in the temple of Olympia.

The Egyptian rule of thumb was easily remembered by the ratios of the different sides of an isosceles triangle formed by a base length 5 and the other two sides of length 8 .

Graphing the equation $x^2 - x - 1 = f(x)$ is easily performed by first completing the square to

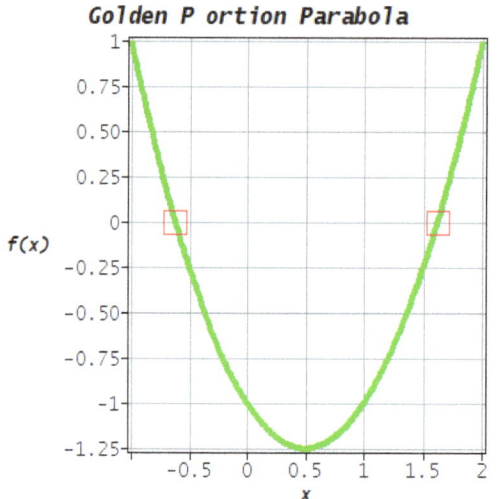

Figure 34

obtain $f(x) = \left(x - \dfrac{1}{2}\right)^2 - \dfrac{5}{4}$. $f(x) = x^2 - x - 1$. Also, construction of the length Phi (ϕ) by compass and ruler can be made with the following steps:

1. Draw segment (line) of length a, $\overline{AB}$.

2. Draw perpendicular line segment at right end of step 1 of length $\dfrac{1}{2}a = \overline{BC}$

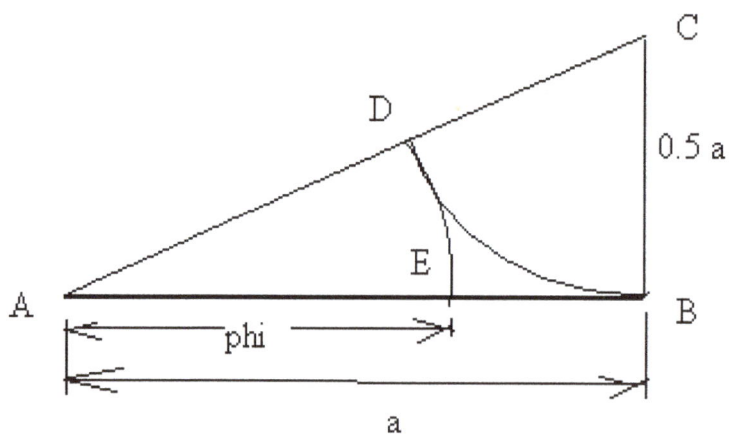

Figure 35

T. Margulies, *Pyramid Geometry Design*

3. Use compass centered at point C to swing arc and radius of length $\overline{BC} = \dfrac{1}{2}a$.

The intersection on $\overline{AC}$ is designated as point D.

4. Using the compass generate an arc of radius $\overline{AD}$ centered at A.

The crossing $\overline{AB}$ is represented as point E.

5. The line segment connecting points A and E is of length *phi*, φ.

The idea of phi includes total balance in architecture design and not just the numerical values. Numerous examples of the Golden ration or its approximation are drawn from nature and architecture. The Greek Parthenon, translated as the *Virgin's place* was constructed during the fifth century B.C. has a side-view with Golden phi length and width.

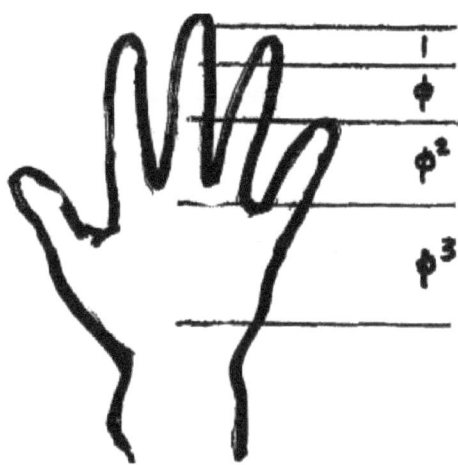

Figure 36

This derivation corresponds to the interpretation of the length to width ratio of line segments which are like the side-lengths of a rectangle, the longer side is approximately 1.61803 multiplied by the shorter side length. An alternate derivation defines a harmonic mean by the geometric proportions, a:b: (a+b), obtains $\dfrac{a-b}{b} = \dfrac{b-(a+b)}{a+b}$ or $\dfrac{a-b}{b} = \dfrac{-a}{a+b}$, so that by multiplying diagonally and dividing by b^2, yields the quadratic equation $x^2+x-1=0$, where $x = \dfrac{b}{a}$. Then solving $x = \dfrac{-1\pm\sqrt{5}}{2}$. The harmonic mean between two numbers is usually presented by a formula

T. Margulies, *Pyramid Geometry Design*

$$\mu_H = 2\frac{ab}{a+b} = \frac{ab}{\frac{1}{2}(a+b)} = \frac{1}{\left(\frac{1a+b}{2\ ab}\right)} = 1/\frac{1}{2}\left(\frac{1}{a} + \frac{1}{b}\right), \text{ or finally} \quad \frac{1}{\mu_H} = \frac{1}{2}\left(\frac{1}{a} + \frac{1}{b}\right)$$

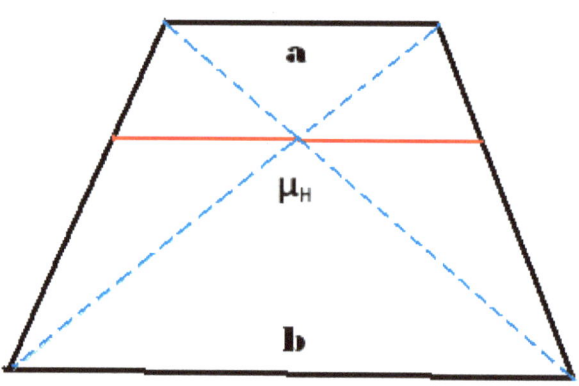

Figure 37

Mathematical Golden Angle

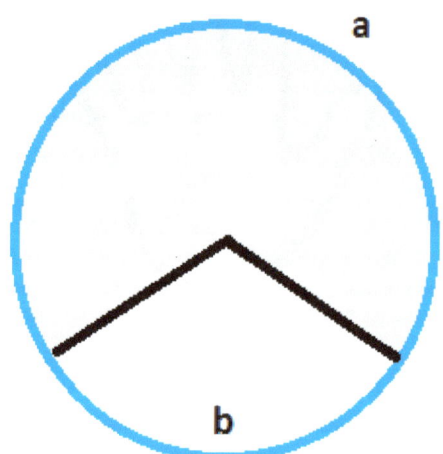

Figure 38

$$\frac{b}{a} = \frac{a}{a+b} \quad , \quad b(a+b) = a^2, \ ab + b^2 = a^2 .$$

Divide by a^2 to obtain
$$\left(\frac{b}{a}\right)^2 - \frac{b}{a} - 1 = 0, \ \ x = \frac{b}{a}$$

$$x^2 - x - 1 = 0, \ \ x = \frac{1+\sqrt{5}}{2} \sim 1.61803 \text{ and } = \frac{1-\sqrt{5}}{2} \sim 0.61803$$

a=222.4922   b=137.5078

T. Margulies, *Pyramid Geometry Design*

For a 360 degrees arc,

137.5degrees is referred as a Golden angle.

Defining a harmonic mean by the geometric proportions, for example 6:4:3, as $\frac{6-4}{6} = \frac{4-3}{3}$.

Likewise a:b: (a+b), obtains $\frac{a-b}{b} = \frac{b-(a+b)}{a+b}$ or $\frac{a-b}{b} = \frac{-a}{a+b}$, so that

by multiplying diagonally and dividing by b^2, yields the quadratic

equation $x^2+x-1=0$, where $x = \frac{b}{a}$.   Then solving $x = \frac{-1\pm\sqrt{5}}{2}$.

Gillings presents a pyramid angle in problem 56 of the *Rhind Mathematical Papyrus*, i.e. φ = arctan(7 palms/ (5 + 1/25) palms) = 54.246° (54°14'46").

Vesica Piscis Golden Phi Calculation

The Vesica Piscis associations with a Mother goddess portrayal.

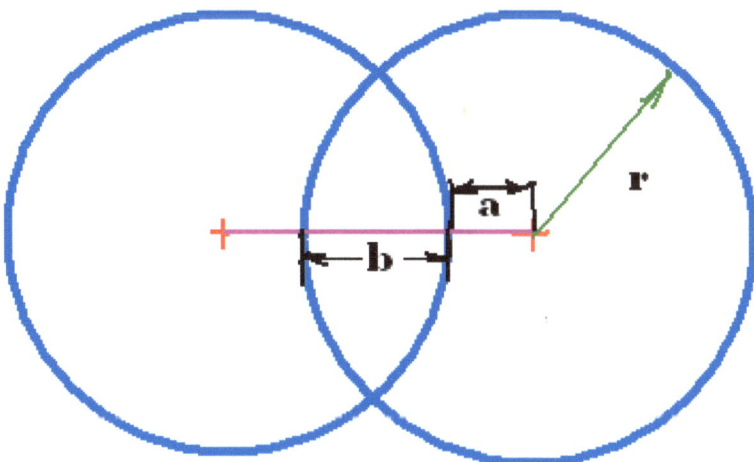

Figure 39

$$\frac{a}{b} = \frac{b}{a+b} \quad b = r - a \quad \frac{a}{r-a} = \frac{r-a}{r} \quad a \cdot r = r^2 - a^2$$

$$r^2 - a \cdot r - a^2 = 0$$

$$\frac{r^2}{a^2} - \frac{r}{a} - 1 = 0 \quad x = \frac{r}{a} \quad x^2 - x - 1 = 0 \quad , \quad x = \frac{1\pm\sqrt{5}}{2}$$

T. Margulies, *Pyramid Geometry Design*

Face and human body measurements display Golden ratios of lengths.

C

B

A

Figure 40

Nefertiti (ca. 1370 BC – ca. 1330 BC) was the Great Royal Wife (chief consort) of the Egyptian Pharaoh Akhenaten. Nefertiti and her husband were known for a religious revolution, in which they worshiped one god only, Aten, or the sun disc. The height of the *cheek-bone* $\frac{A}{B} \sim 0.62$ as well as $\frac{A+B}{C} \sim 0.61$ are Golden Ratios.

Gillings also uncovers a theoretical value of a hypothetical pyramid angle in problem 56 of the Rhind Mathematical Papyrus, i.e. φ = arctan(7 palms/ (5 + 1/25) palms) = 54.246° (54°14′46″). Calculating the volume of a

In a chromatic scale of twelve notes or tones an octave is a doubling of the frequency that would be associated with halving the length of a plucked string. An aesthetic sound to hearing is the musical fifth which is a ratio of string segment lengths of three to two or 3/2 = 1.5 . A major third (5/4) is comprised of a ratio equal to 1.25 . A minor sixth is 8/5 = 1.6 or approximately phi.

A major third is approximately equal to the square root of Phi

$$(\frac{\phi^{0.5}}{major\ third} = \frac{1.272}{1.25} = 1.0176 = 101.76\%).$$

Fibronacci Series And The Golden Ratio

Fibronacci of Pisa (1170-1250 AD) is credited with an amazing series with applications in nature and

an embedded golden phi ratio. The Fibronacci series is formed from the integers by the rule:

T. Margulies, *Pyramid Geometry Design*

let the first two terms be given by 1, 1. Then the next terms are given recursively by summing the previous two terms. The number series becomes 1, 1, 1+1=2, 1+2=3, 2+3=5, 3+5=8, 5+8=13,.... This was first discussed by Leonardo of Pisa. An approximate value of phi is found by dividing one number in the series with its preceding neighbor. These numbers can be used to construct another series where the Fibronacci numbers are both in the numerators and denominators, $\phi^2 = \phi + 1$

$$\frac{1}{1}, \frac{2}{1}, \frac{3}{2}, \frac{5}{3}, \frac{8}{5}, \frac{13}{8}, \frac{21}{13}, ... \quad or \quad write \quad as$$

$$\phi = 1 + \frac{1}{\phi} = 1 + \cfrac{1}{1 + \cfrac{1}{1 + \cfrac{1}{1 + \cfrac{1}{1 + ...}}}},$$

In the limit this approaches phi. Phi may also be written as, $\phi^2 = 1 + \phi$, $\phi = \sqrt{1 + \phi}$ then

$$\phi = \sqrt{1 + \sqrt{1 + \sqrt{1 + \sqrt{1 + \sqrt{1 + ...}}}}}$$

The Fibronacci sequence, $x_n = x_{n-1} + x_{n-2}$ for $n > 2$ where $x_1 = 1 = x_2$.

In a matrix notation, define

$$\underline{Y}_n = \underline{\underline{A}} \, \underline{Y}_{n-1}, \quad \underline{Y}_n = \begin{pmatrix} x_n \\ x_{n-1} \end{pmatrix}, \quad \underline{Y}_{n-1} = \begin{pmatrix} x_{n-1} \\ x_{n-2} \end{pmatrix}, \quad \underline{\underline{A}} = \begin{pmatrix} 1 & 1 \\ 1 & 0 \end{pmatrix}$$

$$\underline{Y}_2 = \begin{pmatrix} x_2 \\ x_1 \end{pmatrix} = \begin{pmatrix} 1 \\ 1 \end{pmatrix}$$

$$\underline{Y}_3 = \underline{\underline{A}} \, \underline{Y}_2, \, \underline{Y}_4 = \underline{\underline{A}} \, \underline{Y}_3 = \underline{\underline{A}}^2 \underline{Y}_2, \, \underline{Y}_5 = \underline{\underline{A}} \, \underline{Y}_4 = \underline{\underline{A}}^3 \underline{Y}_2, \, ... \, \underline{Y}_j = \underline{\underline{A}} \, \underline{Y}_{j-1} = \underline{\underline{A}}^{(j-2)} \underline{Y}_2$$

The eigenvalues are found by solving the determinant equation, $\left| \lambda \underline{\underline{I}} - \underline{\underline{A}} \right| = \begin{vmatrix} \lambda - 1 & 1 \\ 1 & \lambda \end{vmatrix} = 0$ or

$\lambda^2 - \lambda - 1 = 0$. $\lambda_1 = \frac{1+\sqrt{5}}{2} = \phi_1$. $\lambda_2 = \frac{1-\sqrt{5}}{2} = \phi_2$. Note $\phi_1 + \phi_2 = 1$ and $\phi_1 \cdot \phi_2 = -1$.

Real eigenvalues and this transformation matrix exists for this positive definite symmetric matrix $\underline{\underline{A}}$ which is used for diagonalization.

The eigenvectors are obtained by solving for each eigenvalue the equation, $\underline{\underline{A}} \, \underline{v} = \lambda \underline{v}$

$$\begin{pmatrix} 1 & 1 \\ 1 & 0 \end{pmatrix}\begin{pmatrix} v_{11} \\ v_{21} \end{pmatrix} = \phi_1 \begin{pmatrix} v_{11} \\ v_{21} \end{pmatrix}$$ then $v_{11} + v_{21} = \phi_1 v_{11}$ or $v_{11} = \phi_1 v_{21}$. Let $v_{21} = 1$

$$\begin{pmatrix} v_{11} \\ v_{21} \end{pmatrix} = \begin{pmatrix} \phi_1 \\ 1 \end{pmatrix}.$$ Also, $$\begin{pmatrix} 1 & 1 \\ 1 & 0 \end{pmatrix}\begin{pmatrix} v_{12} \\ v_{22} \end{pmatrix} = \phi_2 \begin{pmatrix} v_{12} \\ v_{22} \end{pmatrix}$$

then $v_{12} + v_{22} = \phi_2 v_{12}$ or $v_{12} = \phi_2 v_{22}$. Let $v_{22} = 1$, $$\begin{pmatrix} v_{12} \\ v_{22} \end{pmatrix} = \begin{pmatrix} \phi_2 \\ 1 \end{pmatrix}$$

Let $\underline{\underline{M}} = \begin{pmatrix} \phi_1 & \phi_2 \\ 1 & 1 \end{pmatrix}$ and form its powers, $\underline{\underline{M}}^{(j-2)} \equiv \underline{\underline{T}}$.

Transform the equation, $\underline{Y_j} = \underline{\underline{A}}^{(j-2)}\underline{Y_2}$ by $\underline{Y_j} = \underline{\underline{T}}\,\underline{Z_j}$, so that $\underline{Y_2} = \underline{\underline{T}}\,\underline{Z_2}$

$\underline{Z_j} = \underline{\underline{T}}^{-1}\underline{\underline{A}}^{(j-2)}\underline{\underline{T}}\,\underline{Z_2}$ where $\underline{\underline{T}}^{-1}\underline{\underline{T}} = \underline{\underline{I}}$ for the inverse matrix $\underline{\underline{T}}^{-1}$.

$$\underline{Z_j} = (\underline{\underline{M}}^{(j-2)})^{-1}\underline{\underline{A}}^{(j-2)}\underline{\underline{M}}^{(j-2)}\,\underline{Z_2}$$

$$(\underline{\underline{M}}^{(j-2)})^{-1} = \left(\underline{\underline{M}}^{-1}\right)^{j-2}$$

$$\underline{Z_j} = \left(\underline{\underline{M}}^{-1}\right)^{j-2}\underline{\underline{A}}^{(j-2)}\underline{\underline{M}}^{(j-2)}\,\underline{Z_2}$$

$$\underline{Z_j} = \left(\underline{\underline{M}}^{-1}\underline{\underline{A}}\,\underline{\underline{M}}\right)^{(j-2)}\underline{Z_2}$$

The matrix multiplications were designed to obtain a diagonal matrix of eigenvalues,

$$\underline{\underline{M}}^{-1}\underline{\underline{A}}\,\underline{\underline{M}} = \begin{pmatrix} \lambda_1 & 0 \\ 0 & \lambda_2 \end{pmatrix} \equiv \underline{\underline{D}}$$

$$\underline{Z_j} = \underline{\underline{D}}^{(j-2)}\underline{Z_2} = \begin{pmatrix} (\phi_1)^{j-2} & 0 \\ 0 & (\phi_2)^{j-2} \end{pmatrix}\underline{Z_2}$$

The powers of a diagonal matrix is a matrix is calculated by non-zero elements on the diagonal and zeros elsewhere, with the diagonal elements raised to that power.

On The *Golden* Color Yellow and Portion

Isaac Newton demonstrated that light which was previously believed to be colorless existed as "corpuscles" or particles of light as different colors traveled with different speeds through a prism. Subsequently, Young and Fresnel combined Newton's particle theory with Huygens' wave theory to show that color is the visible manifestation of light's wavelength.

Newton experimented by passing a red color light from one prism through a second prism finding the color unchanged. He concluded that the colors must already be present in the incoming light ;

T. Margulies, *Pyramid Geometry Design*

therefore, a prism does not create colors, but merely separates colors of different wavelengths mixed together in the white light.

Roger Bacon, an English philosopher from the 13th century, postulated, but could not demonstrate, that the colors of a rainbow are due to the reflection and refraction of sunlight through individual raindrops.

Colors of light are decomposed into a threefold blend of primary colors (red, green, and blue-violet). Primary colors for pigments are based on reflections of light. That is, inks, caulks, crayons, paints and pencils would have primaries red, yellow, and blue. The ink on the page appears as black because all the light is absorbed or "subtracted" and what remains is an absence of color, no color. Scientists have found in analyzing light, the different color sensations are due to varying wavelengths of light to which the eye is sensitive. Considering the rainbow, red light has the largest wavelength, orange the next largest, and so on down to yellow, green, indigo, and finally violet with the shortest wavelength. This spectrum of colors can be obtained using a prism to separate the light into different wavelengths. The following table summarizes approximate wavelengths and frequencies of several colors of light. The wavelength units are angstroms (or 10^{-10} meters).

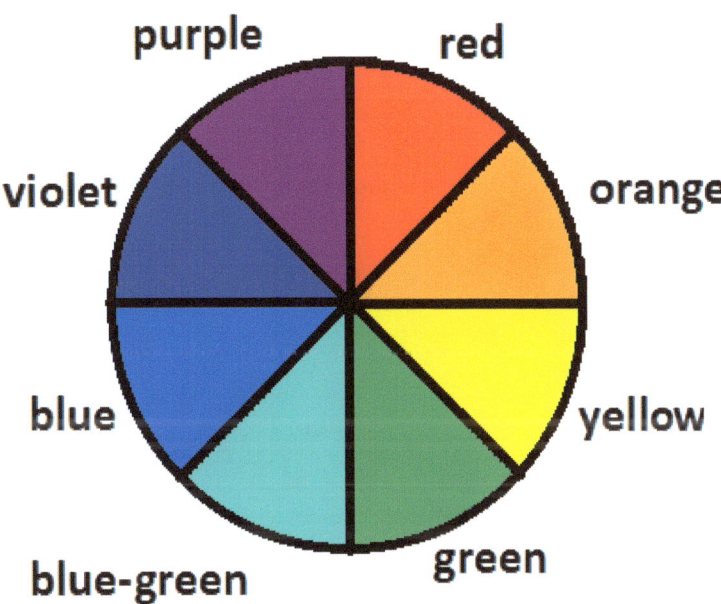

Figure 41

Table 14

Color	Wavelength (Angstroms)	Frequency (Waves / sec)
Red	7100	$4.23 \ 10^{14}$
Orange	6200	$4.83 \ 10^{14}$
Yellow	5700	$5.25 \ 10^{14}$
Green	5200	$5.76 \ 10^{14}$
Blue	4700	$6.39 \ 10^{14}$
Indigo	4300	$6.98 \ 10^{14}$
Violet	4100	$7.32 \ 10^{14}$

It is noted that ratio of the red-blue part of the spectrum to the red – yellow is approximately

$\dfrac{7100 - 4700}{7100 - 5700} \approx 1.71$, near a golden phi for the wavelengths corresponding to the color of yellow

light as the "golden portion". Also, for a blue-green average of 4950 angstroms,

the red- [blue-green] portion of the visible spectrum to red-yellow is about $\dfrac{7100 - 4950}{7100 - 5700} \approx 1.54$.

A ratio of red through to an average blue-indigo color to a red- yellow portion is $\dfrac{7100 - 4500}{7100 - 5700} \approx 1.85$.

Calculations for the human visual system reveal the following:

Table 15

blue	green	red
444.4	526.3	645.2
	81.9	118.9
	(g - b)/(r - g) =	0.6888141

The bald eagle has pale yellow or brownish eye-color and density of rods and depth of fovea achieves an estimated 20/5 visual acuity.

Figure 42

Sphinx Paws Discovery at Tel Hazor

A stick or pole would translate as *word*, and several sticks would mean the plural form, words. The important work of writing by scribes would start with a prayer,

O Thot [Master of hieroglyphs] deliver me from
useless speech. Be behind me in the morning. Come, you who are divine
speech. You are a sweet fountain for the thirsty desert traveler. It is closed
for the garrulous and open for the silent. [Sallier Papyrus 1,8,2-6]

T. Margulies, *Pyramid Geometry Design*

Figure 43: Sphinx Paws Discovery at Tel Hazor, Israel (Josua Talbot presenter [16.07.13 Ägyptologie DIE WELT]).

A possible Interpretation of the hieroglyphs follows but awaits closer examination and verification. The interpretation of this 4500 year old archaeological discovery attributed to Pharaoh Mycerinus who built the smallest of the three pyramids at Giza, southwest of Cairo, Egypt. This granite fragment was found north of the Sea of Galilee and announced by Hebrew University archaeologists Amnon Ben-Tor and Dr. Sharon Zuckerman. to Pharaoh Mycerinus [Menkaure] who probably ruled when Moses led the Israelites out of Egypt. Menkaure is divine is the name of the pyramid at Giza, The Netjer-er-Menkaure, or eternal like the souls of Ra. The third of the pyramids that he built followed the first two larger constructed by his grandfather and father, respectively. Herodotus believed he was a benevolent ruler. Accounts on this Old Kingdom time of Egypt depict that his mind was disturbed since prophesies of 150 years of tyranny surrounded his relatively reign.

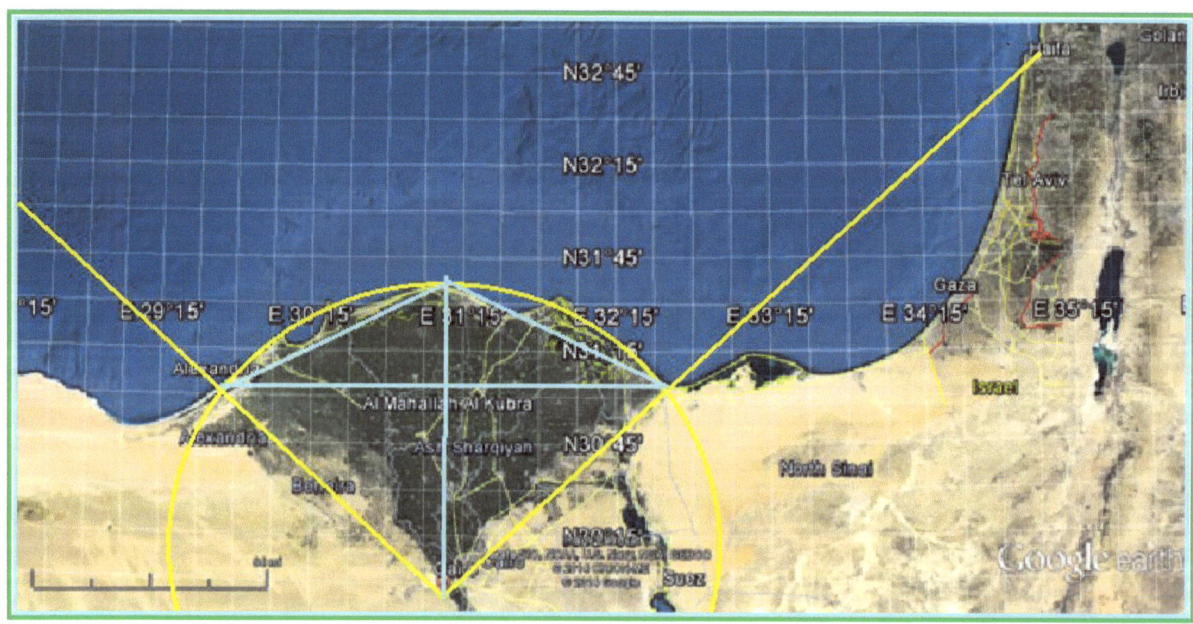

Figure 44

Table 16

	Abide (Abiding-continue without change, believe)	Symmetry line
	Abide Three men (chest and arms)	Priests, Pharaohs *In greater arms; afterlife*
	Lord	
	Pigeon/Falcon	Horus right eye
	Receive, take	Divine souls
	Sun	bread
	radiance	Company of 9 gods
	Garland (flowers tied in a bundle)	
	Milk	
	Mouth	
	Perfect health	Twenty
	Beauty	Pyramid Phi design Architects rule 8/5 isosceles triangle
	Life	
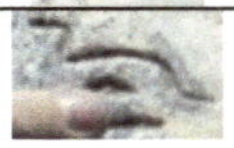	Forever, everlasting	Passageway aligns to Orion/Draconis

Concluding Remarks

 A list summarizing findings and remarks follow.

   Earth has many archaeological mysteries that delight viewers' imaginations and challenge scientists for rational explanation. Several are highlighted to arouse interest, speculation, and perhaps some investigation.

The Golden Phi was incorporated in the designs of pyramids at Giza. The possible connection to Nefertiti's face appears worthy of consideration for exhibiting a most beautiful form.

T. Margulies, *Pyramid Geometry Design*

The pyramid design mathematics for calculating the volume of an oblique or a regular pyramid partitioned into calculable shapes has been presented. The same volume formula which applies to both and the one-third factor is easily shown.

The emphasis of light, especially that from Stars and the global practice of pyramid building encircling Earth is a robust ubiquitous insight.

A Nazca line drawing expresses a sign of pyramid light, perhaps eclipse, global marking arrangement.

The nonlinear spatial arrangement of Great Pyramids at Giza may be explained by volume, or directly proportional height, correlations with the light from the three belt Stars of the Orion Constellation or stars from the Cygnus Constellation.

In closing, a historical comment is made on the United States Great Seal which was adopted by Congress in 1782 after several design proposals. The seal's reverse side contains a requested truncated, or expressed *unfinished pyramid*. A proposed sketch by Hopkinson is depicted on the left and the final choice on the right in the figure. At the top is the Eye of Providence with the Latin motto *Annuit Coeptis* in the sky above translated as, *It* [Eye of Providence] *is favorable to our undertakings* or *He favors our undertakings*. The scroll reads *Novus Ordo Seclorum* which is Latin for *New Order of the Ages*.

Figure 45

Acknowledgement:

The Golden phi ratio of the visible light spectrum of humans as yellowish was suggested by Dr. Peter Lennie.

T. Margulies, *Pyramid Geometry Design*

References:

Allred, Sarah R. and D. H. Brainard, Contrast, constancy and measurements of perceived lightness under parametric manipulation of surface slant and surface reflectance, *JOSA* 26(4) pp. 949-961 (2009).

Anton, Howard, *Elementary Linear Algebra*, Wiley (2013).

Balkich, Michael, *The Cambridge Planetary Handbook*, Cambridge (2000).

Bauval, R.G., A Master Plan for Three Pyramids of Giza Based on the Configuration of Three Stars of the Belt of Orion, Discussions in Egyptology, 13, 7 (1989).

Brummelen, Glen van, *Heavenly Mathematics: The Forgotten Art of Spherical Trigonometry*, Princeton University Presss (2012).

Davidson, D., *Great Pyramid: It's Divine Message*, Kessinger (1992).

Davidson, D. and H. Aldersmith, *The Great Pyramid Its Divine Purpose, vol. I: Pyramid Records*, Fourth Edition, Williams and Norgate, Ltd. (1927).

Friberg, Joran, *A Remarkable Collection of Babylonian Math Texts*, Springer (2007).

Gillings, 1972 *Mathematics in the Time of the Pharaohs*, MIT Press.

Green, David M. and John A. Swets, *Signal Detection theory and Psychophysics*, Peninsula Publications (1989).

Griffiths, Francis Ll, *A Collection of Hieroglyphics*, Egyptian Exploration Fund (1898).

Hecht, Seli, S. Shlaer, and M. H. Pirenne, Energy, Quanta, and Vision, *J. Gen. Physiol.*, vol. 25(6) pp. 819-840 (1942).

Hohenkerk, C.Y. and A.T. Sinclair, *The Computation of Angular Atmospheric refraction at Large Zenith Angles*, Nautical Almanac Office Technical Note, 63, United Kingdom Hydrographic Office (1985).

Huntley, H.E. ,1970 *The Divine Proportion*, Dover.

Landau, L.D. and E. M. Lifshitz, *Mechanics, Volume I (Course of Theoretical Physics)*, Butterworth-Heineamnn (1975).

Larson, Ron and David C. Falvo, *Elementary Linear Algebra*, Houghtin-Mifflin-Harcourt (2009).

Lehner, Mark and Richard H. Wilkinson, *Complete Pyramids: Solving the Ancient Mysteries*, Thames & Hudson (2002).

Livio, Mario, The *Golden ratio: The Story of Phi, the World's Most Astonishing Number*, Broadway Books (2003).

T. Margulies, *Pyramid Geometry Design*

Lockyer, J. Norman, *Dawn of Astronomy*, Kessinger.

Margulies, Timothy S., Margulies, *Brief Mathematics: Patterns, Calendars, Pyramids, and Vision with Golden Phi*, 2014, Create Space, ISBN-13: 978-1495461088.

Pacioli, Luca., 1494 *De divina proportione (On the Divine Proportion)*, *Summa de arithmetica, geometria, proportionie proportionalita (Everything about Arithmetic, Geometry, and Proportions)*.

Pallett, Pamela M., Stephen Link, and Kang Lee, New "Golden Ratios" for Facial Beauty, *Vision Res.* 2010 Jan 25: 50(2): 149.

Peebles, P.J.E., *Principles of Physical Cosmology*, Princeton University Press (1993).

Rossing, Thomas and C. Chiaverina, 1999 *Light Science: Physics and Visual Arts*, New York, Springer.

Saaty, Thomas L., Multicriteria Decisionmaking: The *Analytic Hierarchy Process,* RWS (1990).

Shevell , Steven K., *The Science of Vision* (2003).

Smith, Robert T. and Roland B. Minton, *Calculus Early Transcendental Functions*, Third Edition McGraw Hill (2007).

Stevens, S. S., On the Psychophysical law, *Psychology Review* 64(3) 153-181 (1957).

Tassoul, Jean Louis and Monique, *A Concise History of Solar and Stellar Physics*, Princeton (2004).

Tompkins, Peter, *Secrets of the Great pyramid: Two Thousand Years of Adventures and Mysteries*, BBS (1997).

Wheeler, Noel F., Pyramids and their purpose, Antiquity, vols. I, II, II, pp. 5-21. 161-, 292-304 (1935).

Wyszecki, G. and Stiles, W.S., *Color Science: Concepts and Methods, Quantitative Data and Formulae*, 2nd Ed., John Wiley & Sons, New York 1982

Appendices:

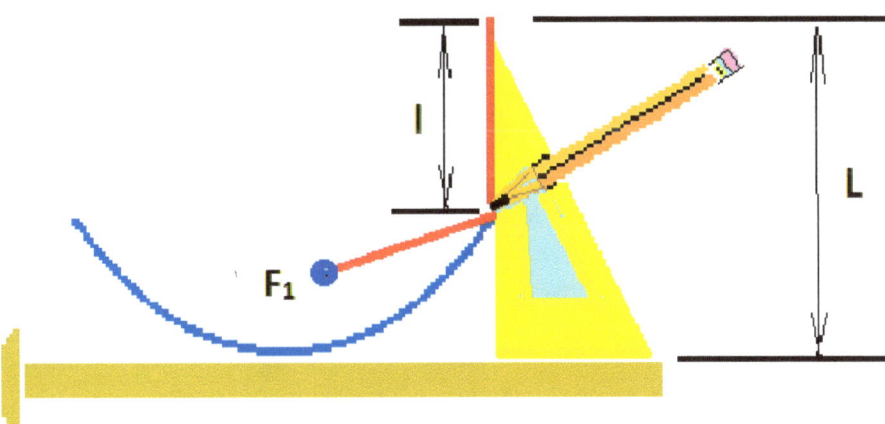

Figure A-1

Parabola: For a string of length y, the distance from the focus point to the pencil tip is $y - l$ and the

Distance from the top of the triangle where the string is also pinned is $L - l$ is the perpendicular

distance to the directrix (T square) $y - l = L - l$ for the string length is $y = L$.

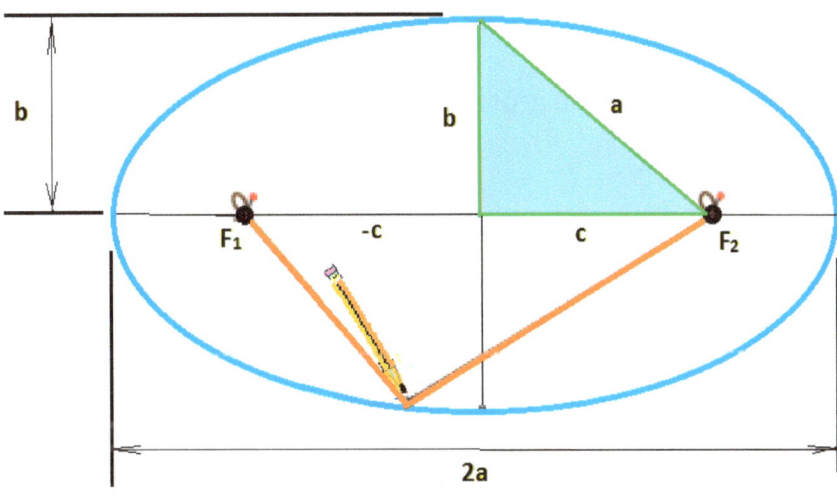

Figure A-2

Ellipse: The distances of a point on the curve to the foci is the same constant string length.

The foci are the pinned ends of the string. The right triangle relationships between the semi-major axis a and semi-minor axis b with foci at distances c and –c for an ellipse centered at the origin are shown.

Ellipse parameter relationships include: eccentricity $e = \frac{c}{a}$, and $b^2 + c^2 = a^2$.

T. Margulies, *Pyramid Geometry Design*

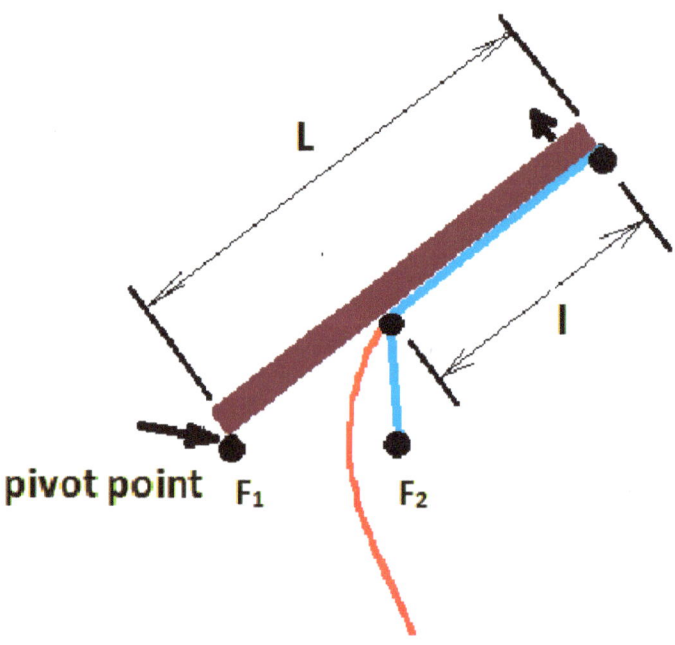

Figure A-3

Hyperbola: The difference of the distances from a point on the curve to the two foci is constant. The pencil is a distance $(L - l)$ from the bar tip focus F_1 and $(l_S - l)$ is the distance to F_2

for string length l_S . Then, $(L - l) - (l_S - l) = constant = L - l_S$

Babylonian mathematics included geometrical problems addressing areas, for example,

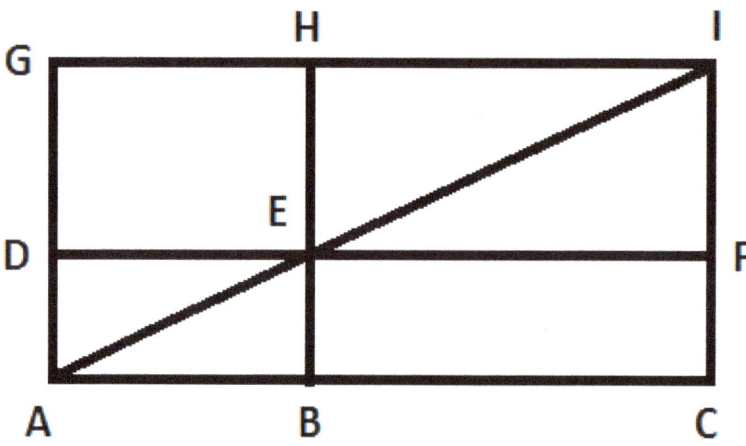

Figure A-4

Area A_T of right triangle given by base GA and height GHI is $A_T = \frac{1}{2}\, \overline{GA} \cdot \overline{GI}$

T. Margulies, *Pyramid Geometry Design*

Area of Rectangle is given as the length and width product $\overline{GA} \cdot \overline{AC}$

Area A_F of AEHGDA is $A_F = \frac{1}{2}(\overline{GA} + \overline{HE}) \cdot \overline{GH}$

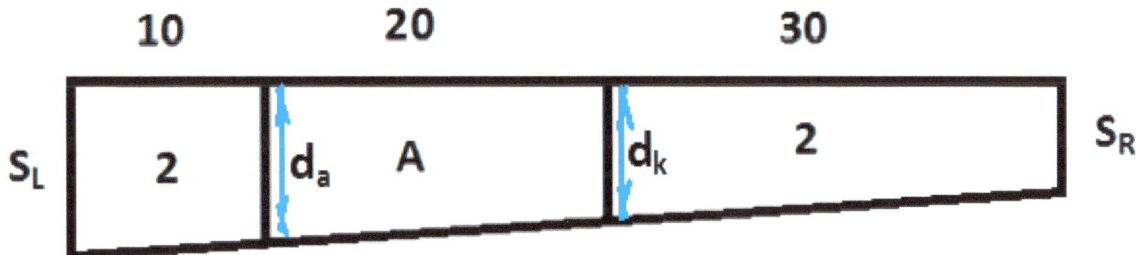

Figure A-5

The Babylonian mathematics presented this problem that used numerical values for five parameters with five variables to be solved by a system of equations. The actual solution method appears unknown and probably by trial and error for integer solutions. Let,

$$S_L + d_a = 24$$

$$A = 20\frac{(d_a + d_k)}{2}$$

$$d_k + S_R = 8$$

$$d_k = d_a - 2(S_L - d_a)$$

$$S_R = d_k - 3(S_L - d_a)$$

Using modern matrix vector notation, $\underline{\underline{A}}\,\underline{x} = \underline{b}$ and solving $\underline{x} = \underline{\underline{A}}^{-1}\underline{b}$ where $\underline{\underline{A}}\,\underline{\underline{A}}^{-1} = \underline{\underline{I}}$ and Rank$\left(\underline{\underline{A}}\right) = 5$

$$\begin{pmatrix} 1 & 0 & 1 & 0 & 0 \\ 0 & 0 & 10 & 10 & -1 \\ 0 & 1 & 0 & 1 & 0 \\ 2 & 0 & -3 & 1 & 0 \\ -3 & -1 & 3 & 1 & 0 \end{pmatrix} \begin{pmatrix} S_L \\ S_R \\ d_a \\ d_k \\ A \end{pmatrix} = \begin{pmatrix} 24 \\ 0 \\ 8 \\ 0 \\ 0 \end{pmatrix}, \underline{x} = \begin{pmatrix} S_L \\ S_R \\ d_a \\ d_k \\ A \end{pmatrix}, \underline{b} = \begin{pmatrix} 24 \\ 0 \\ 8 \\ 0 \\ 0 \end{pmatrix}$$

T. Margulies, *Pyramid Geometry Design*

$$
\underline{\underline{A}} = \begin{pmatrix} 1 & 0 & 1 & 0 & 0 \\ 0 & 0 & 10 & 10 & -1 \\ 0 & 1 & 0 & 1 & 0 \\ 2 & 0 & -3 & 1 & 0 \\ -3 & -1 & 3 & 1 & 0 \end{pmatrix}, \quad \underline{\underline{A}}^{-1} = \begin{pmatrix} 0.5625 & 0 & -0.0625 & 0.125 & -0.0625 \\ -0.1875 & 0 & 0.6875 & -0.375 & -0.3125 \\ 0.4375 & 0 & 0.0625 & -0.125 & 0.0625 \\ 0.1875 & 0 & 0.3125 & 0.375 & 0.3125 \\ 6.25 & -1 & 3.75 & 2.5 & 3.75 \end{pmatrix}
$$

Using Microsoft Excel to find the inverse matrix and perform the matrix multiplication,

$$
\begin{pmatrix} S_L \\ S_R \\ d_a \\ d_k \\ A \end{pmatrix} = \begin{pmatrix} 13 \\ 1 \\ 11 \\ 7 \\ 180 \end{pmatrix}
$$

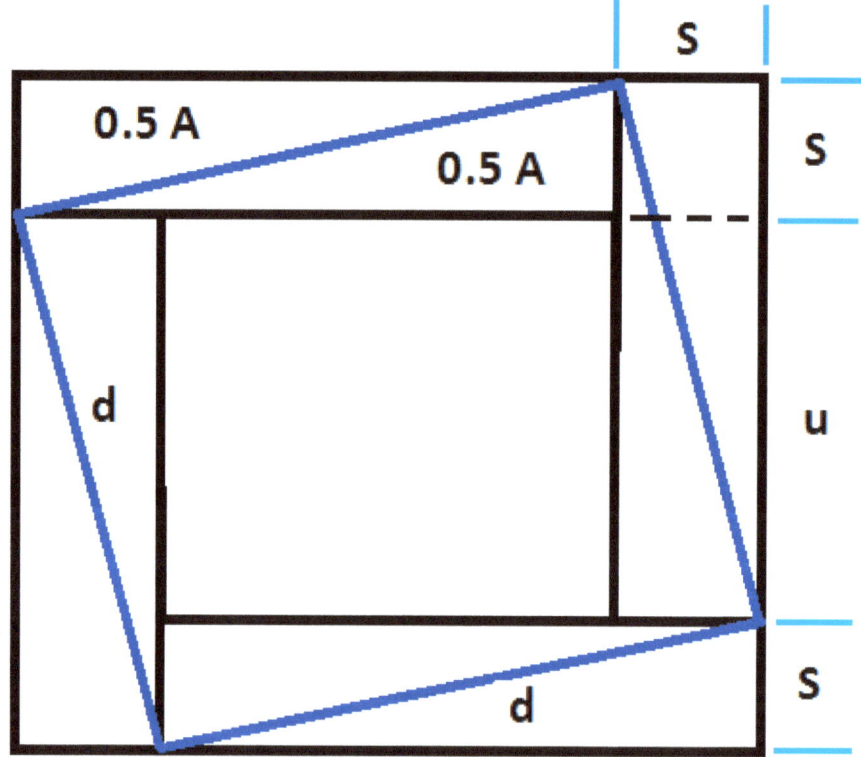

Figure A-6

The areas can be found, for example, by, $\quad u^2 + 4s(u + s)$

Also, $d^2 + 4 \cdot [0.5A] = d^2 + 2A$ and $\quad A = s(u + s) = su + s^2$; equating

$$u^2 + 4su + 4s^2 = d^2 + 2A$$

$$u^2 + 2su + 2s^2 = d^2$$

$$\sqrt{u^2 + 2s(u + s)} = d$$

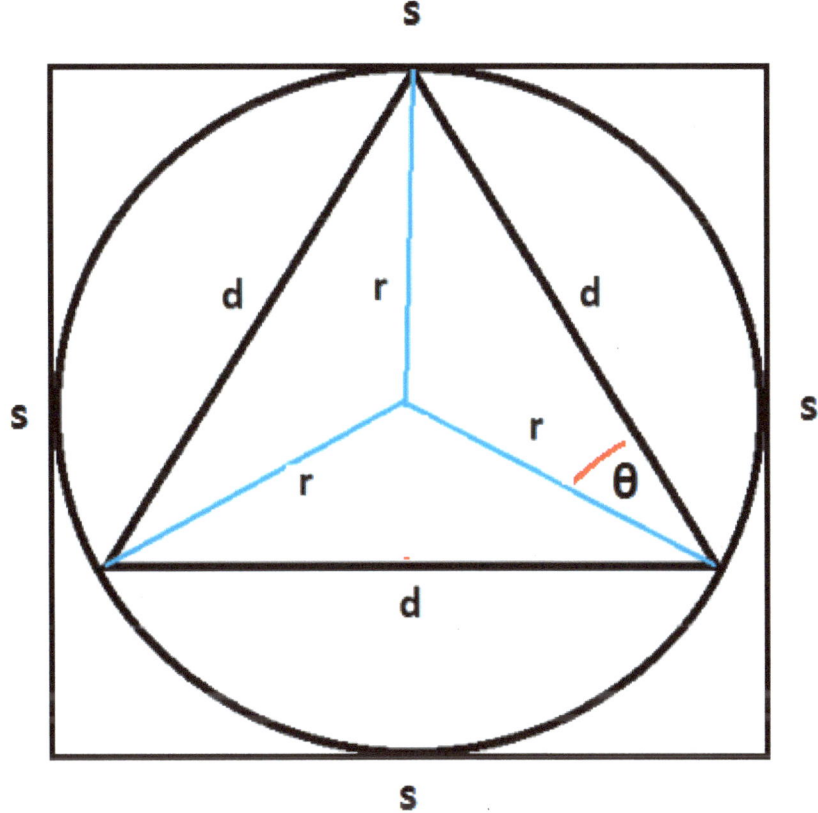

Figure A-7

$$A_{Square} = s^2, \; A_{Circle} = \pi r^2, \; A_{Triangle} = 3dr \cdot sin(\theta)$$

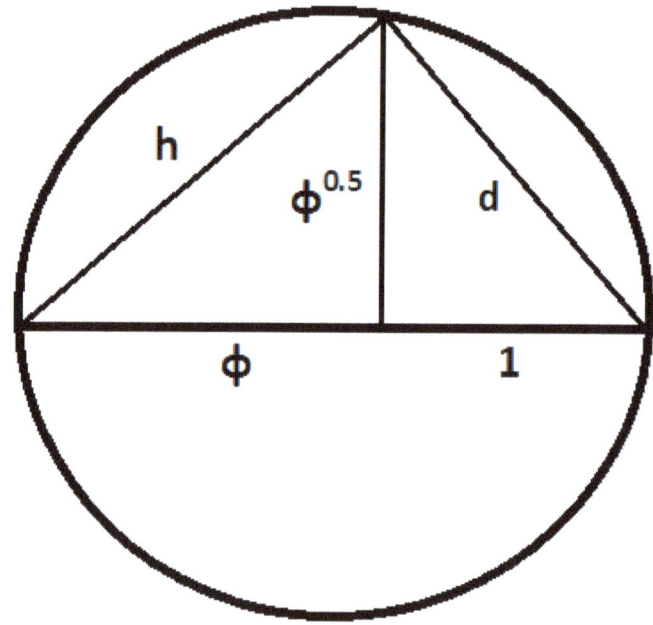

Figure A-8

$h^2 = \phi^2 + \phi$ where the Golden section satisfies

$\phi^2 = \phi + 1$ then, $h^2 = 2\phi + 1$ and $h = \sqrt{2\phi + 1}$ and $\phi + 1 = d^2$; however, $\phi^2 = \phi + 1$

Implies $\phi^2 = d^2$ where $\phi = d$ considering a positive case only.

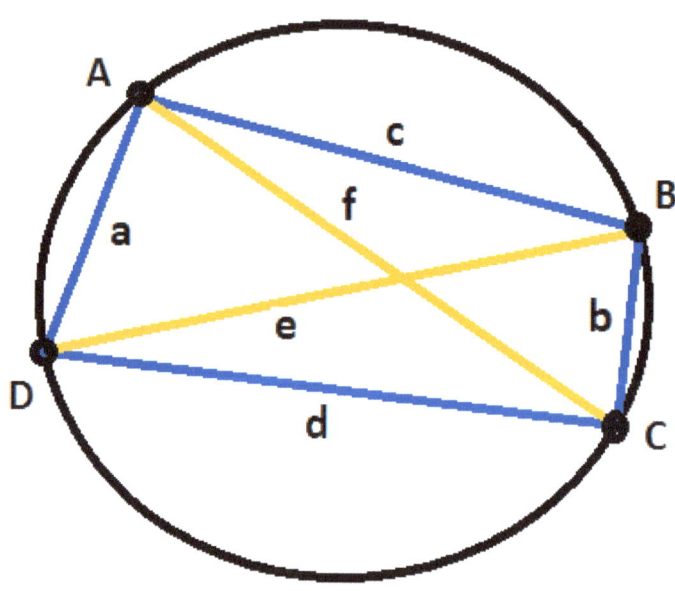

Figure A-9

Chords of a circle forming a quadrilateral ABCD with its diagonals drawn, $a \cdot b + c \cdot d = e \cdot f$

For the special case, $a = b, \ c = d, \ e = f : \quad a^2 + b^2 = e^2$

Part of a regular inscribed polygon forms a kite-like shape with apotherm a, and sagitta, $b = r - a$

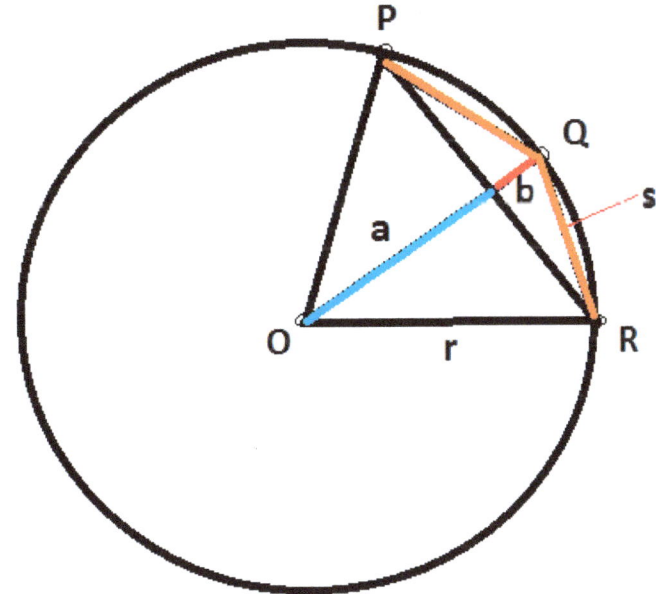

Figure A-10

$$a^2 = r^2 - \left(\frac{c}{2}\right)^2 , \text{ where } c = \overline{PR} \text{ and } s^2 = b^2 + \left(\frac{c}{2}\right)^2 \text{ or } s^2 = 2rb.$$

Power of a Point

In 1826 Jacob Steiner referred to the relationship of a point with two intersecting lines to a circle as "power". The following addresses cases for secants, chords, and tangents.

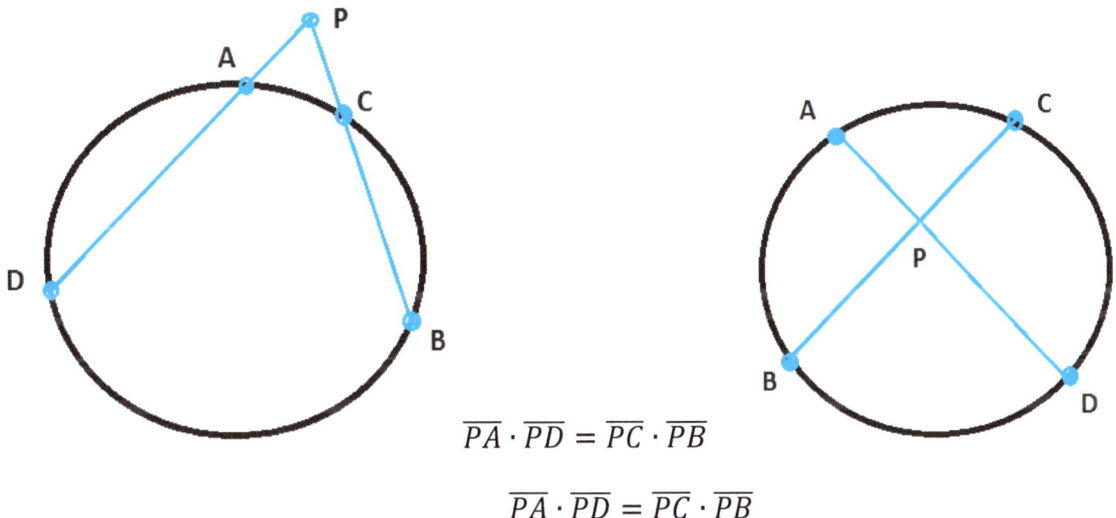

$$\overline{PA} \cdot \overline{PD} = \overline{PC} \cdot \overline{PB}$$

$$\overline{PA} \cdot \overline{PD} = \overline{PC} \cdot \overline{PB}$$

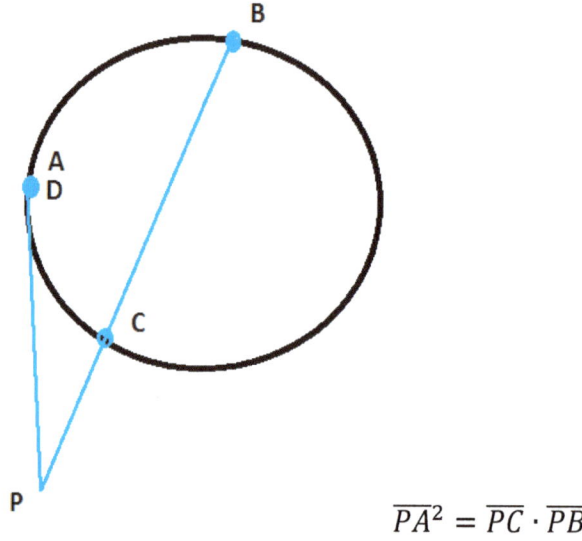

$$\overline{PA}^2 = \overline{PC} \cdot \overline{PB}$$

Figure A-11

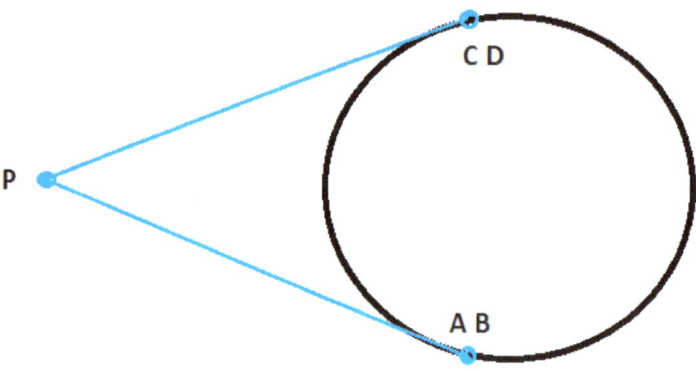

Figure A-12

These line segments are tangent to the circle

$$\overline{PA} = \overline{PC}$$

T. Margulies, *Pyramid Geometry Design*

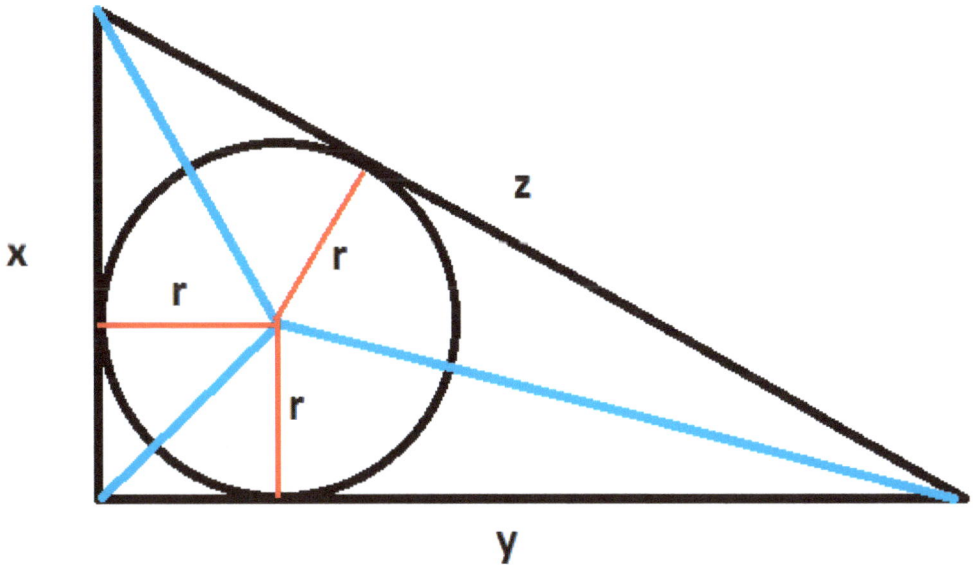

Figure A-13

$$Area = \frac{xy}{2} = \frac{rx}{2} + \frac{ry}{2} + \frac{rz}{2} = \frac{r}{2}(x + y + z) \text{ and solving} \qquad r = \frac{xy}{x+y+z}$$

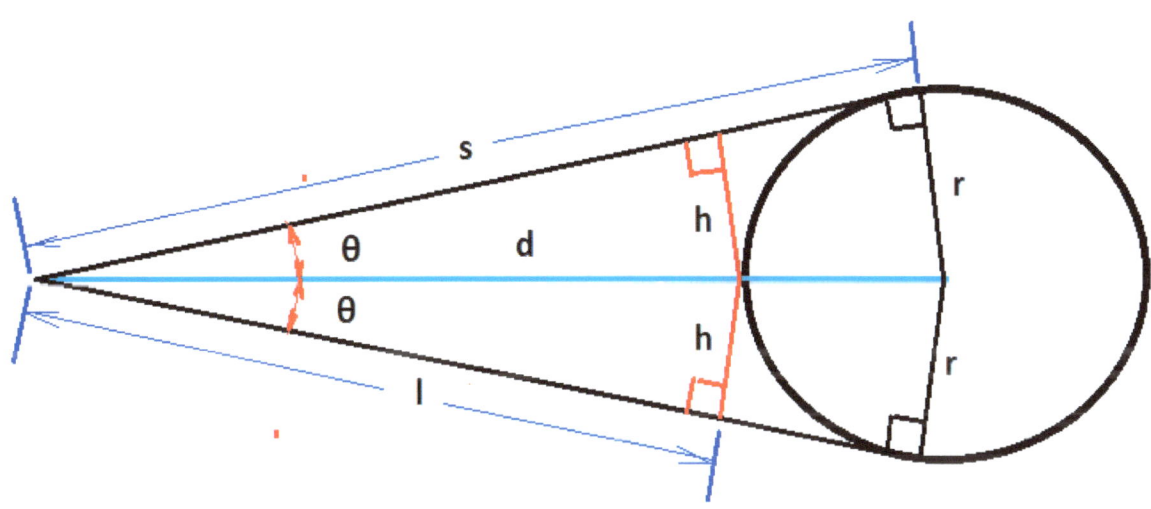

Figure A-14

Right triangle with six trigonometric functions defined below,

T. Margulies, *Pyramid Geometry Design*

$$sine(\theta) = \frac{r}{d}$$

$$cosine(\theta) = \frac{s}{d}$$

$$tangent(\theta) = \frac{r}{s} \quad \text{"Seked function"}$$

$$cotangent(\theta) = \frac{s}{r}$$

$$secant(\theta) = \frac{d}{s} = \frac{1}{cosine(\theta)}$$

$$cosecant(\theta) = \frac{d}{r} = \frac{1}{sine(\theta)}$$

The distance and radius to a celestial body can be estimated using trigonometry, similarity as a *division of figures*, and the Pythagorus theorem.

$$r^2 + s^2 = d^2$$

Euler's Trigonometric Identity

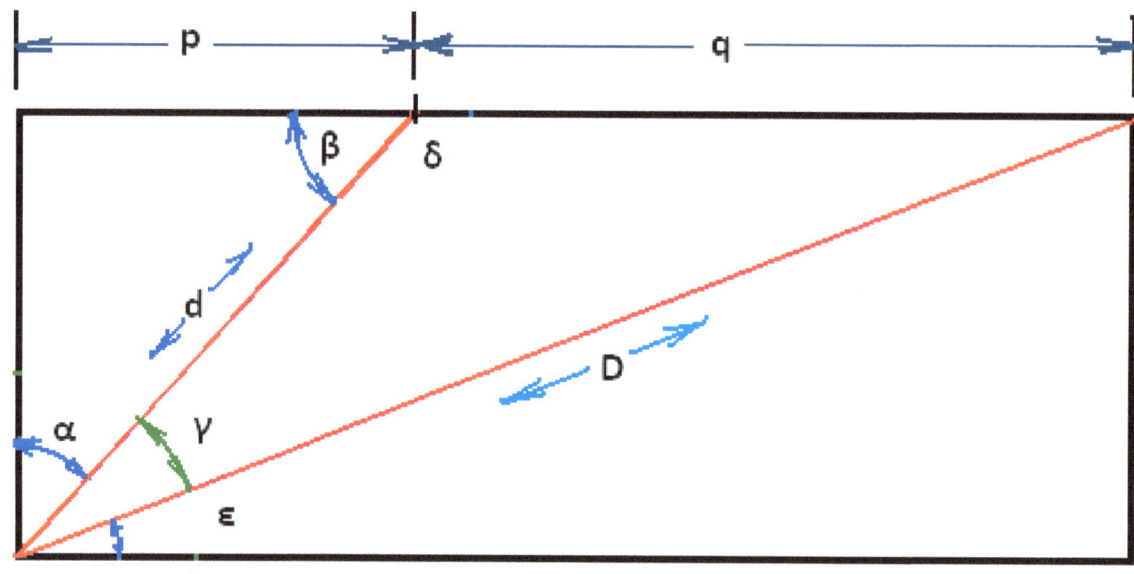

Figure A-15

Using Trigonometry definitions $\boldsymbol{\beta = tan^{-1}\left(\frac{1}{p}\right)}$,

$\boldsymbol{\varepsilon = tan^{-1}\left(\frac{1}{p+q}\right)}$, $\boldsymbol{tan(\gamma) = \frac{sin(\gamma)}{cos(\gamma)}}$.

The Law of Sines, $\qquad \frac{q}{sin(\gamma)} = \frac{D}{sin(\delta)}$, $\quad \frac{q}{sin(\gamma)} = \frac{D}{sin(\beta)}$,

$\frac{q \cdot sin(\beta)}{D} = sin(\gamma)$ where $\delta = \pi - \beta$

The Law of Cosines, $\quad \boldsymbol{q^2 = d^2 + D^2 - 2 \cdot dD \cdot cos(\gamma)}$ or

$$cos(\gamma) = \frac{d^2 + D^2 - q^2}{2 \cdot dD}$$

and Pythagorean Theorem derives the arctangent formula attributed

to Euler.

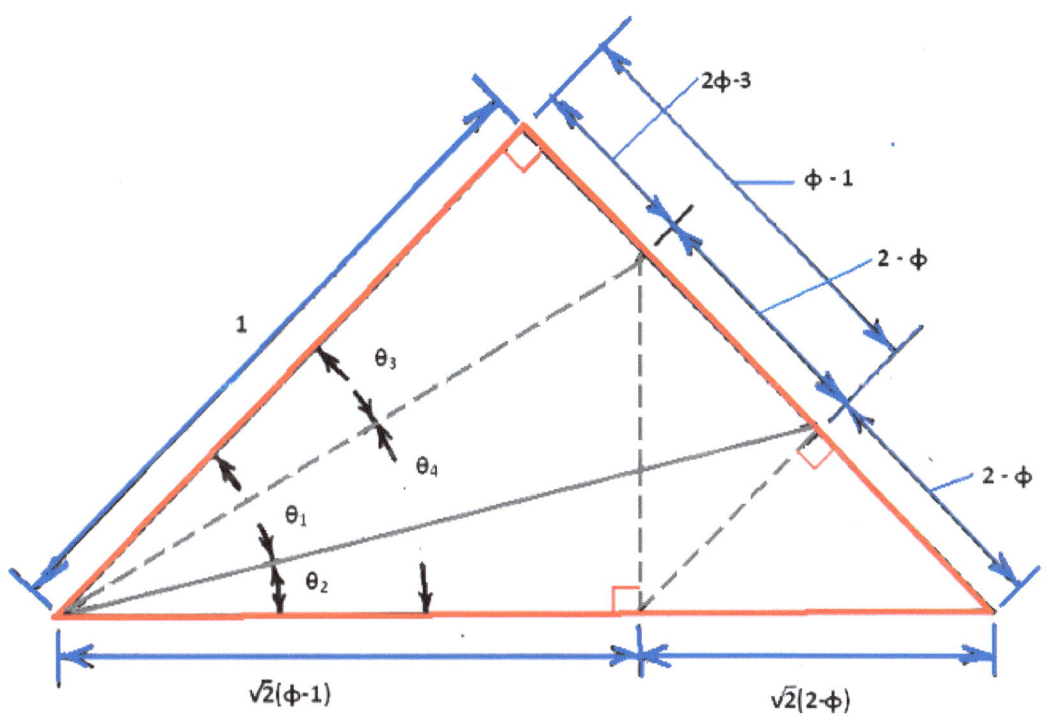

Figure A-16

T. Margulies, *Pyramid Geometry Design*

$$d^2 = p^2 + 1 \text{ and } D^2 = (p + q)^2 + 1.$$

That is,

$$\tan^{-1}\left(\frac{1}{p}\right) = \tan^{-1}\left(\frac{1}{p+q}\right) + \tan^{-1}\left(\frac{q}{p^2 + pq + q^2}\right)$$

Several series representations for Pi can be made. For example,

$$\frac{\pi}{4} = \sum_{k=1}^{\infty} \tan^{-1}\left(\frac{1}{F_{2k+1}}\right)$$

where F_k represent the Fibonacci numbers ($F_2 = 1$, $F_3 = 2$, $F_1 = 3$, $F_1 = 5$, …)

and in general, the n-th number is calculated by $F_n = F_{n-1} + F_{n-2}$

$F_0 = 0$, $F_1 = 1$, $F_2 = 1$, $F_3 = 2$, $F_4 = 3$, $F_5 = 5$, $F_6 = 8$, $F_7 = 13$,…

Alternative series representations such as Chebyshev can also be

made. $\frac{\pi}{4} = \tan^{-1}(1) = \tan^{-1}(1/\phi) + \tan^{-1}(1/\phi^3)$

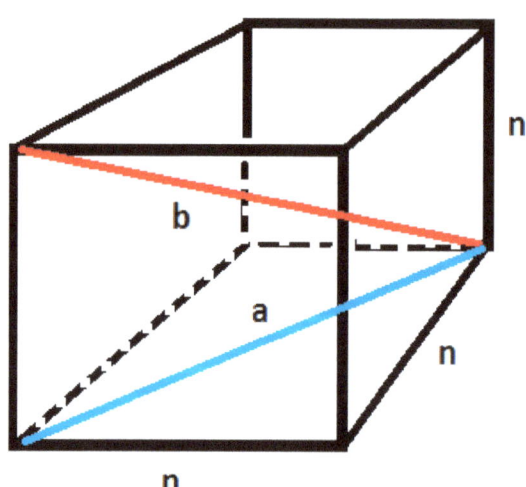

Figure A-17

$$a = \sqrt{2}n \qquad b = \sqrt{3}n$$

For $n = 1$, $a = \sqrt{2}$, $b = \sqrt{3}$

T. Margulies, *Pyramid Geometry Design*

Cube Comprised of Six Pyramids

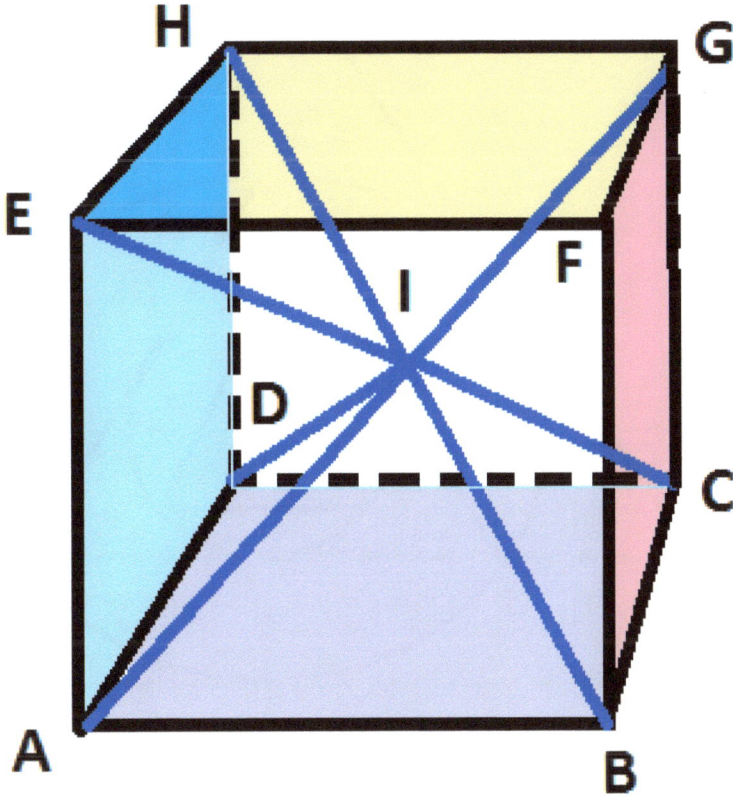

Figure A-18

Let each edge side-length be unity,

$$\overline{AB} = \overline{BC} = \overline{CD} = \overline{AD} = \overline{AE} = \overline{DH} = \overline{AE} - \overline{EH} = \overline{EF} = \overline{FG} = \overline{GH} = \overline{CG} = \overline{BF} = 1$$

Then the volume of the cube is one which is parceled into six rectangular base pyramids that have

point I as their apex. Equating the square base area of one multiplied by the height of one-half and a

variable , c , forms an equation,

$$V_{cube} = 1 = 6\,V_{pyramid} = 6 \cdot c(base \cdot height) = 6 \cdot c\left(1 \cdot \frac{1}{2}\right)$$

Solving , $1 = 3c$ or $c = \frac{1}{3}$.

$$V_{pyramid} = \frac{1}{3}(base \cdot height)$$

T. Margulies, *Pyramid Geometry Design*

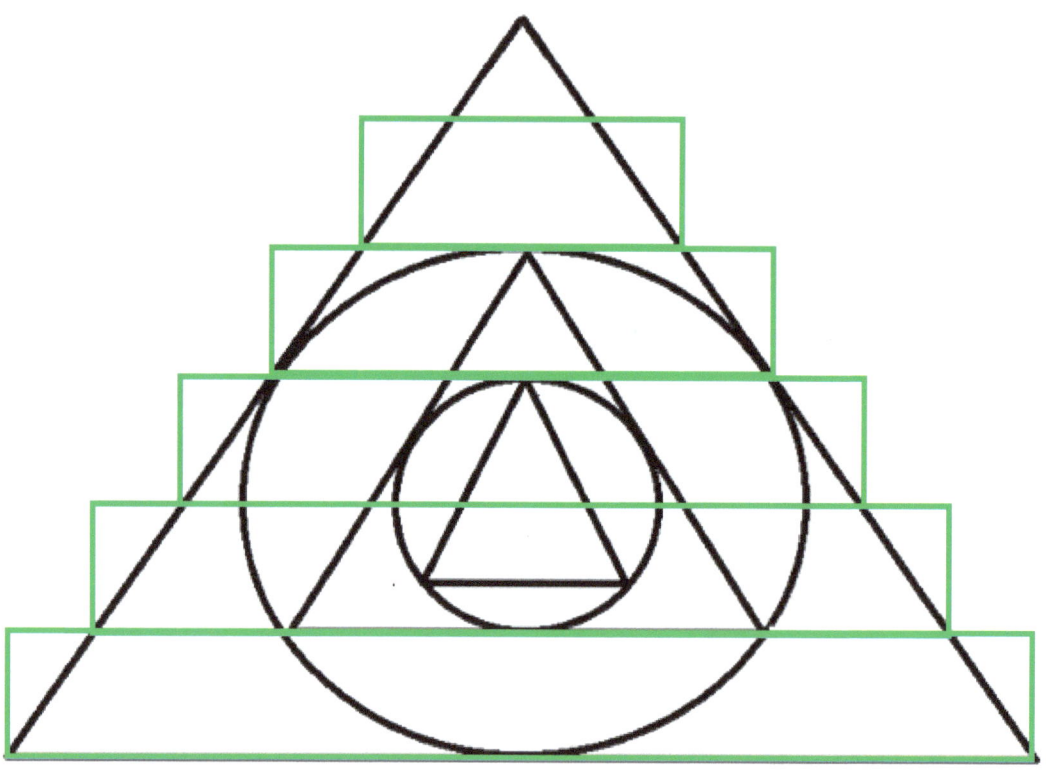

Figure A-19

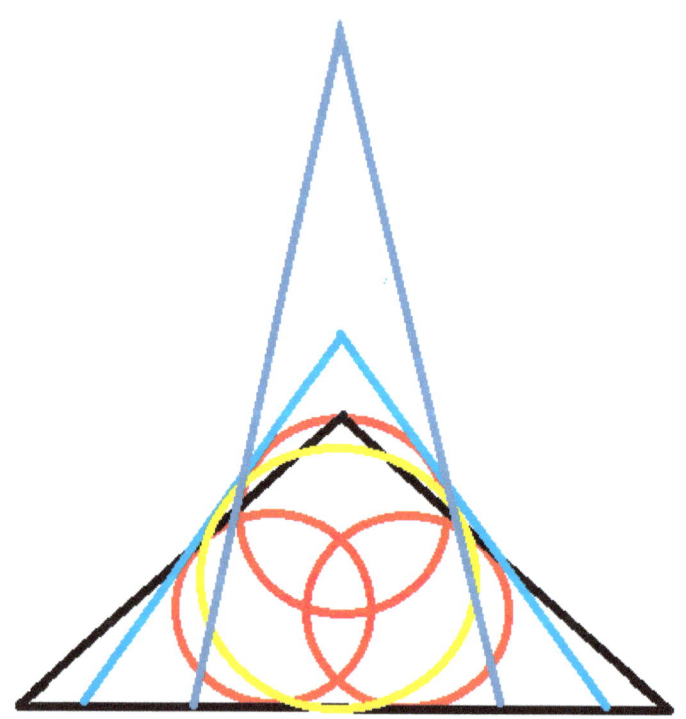

Figure A-20

T. Margulies, *Pyramid Geometry Design*

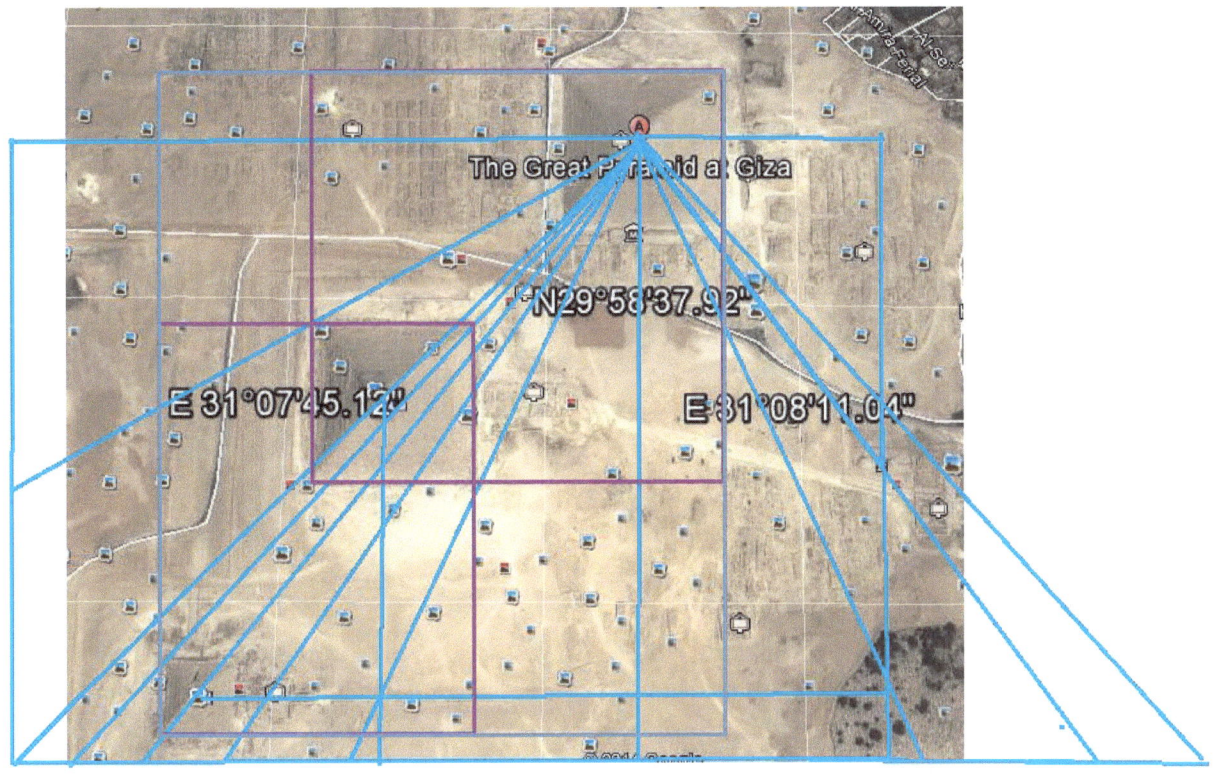

Figure A-21

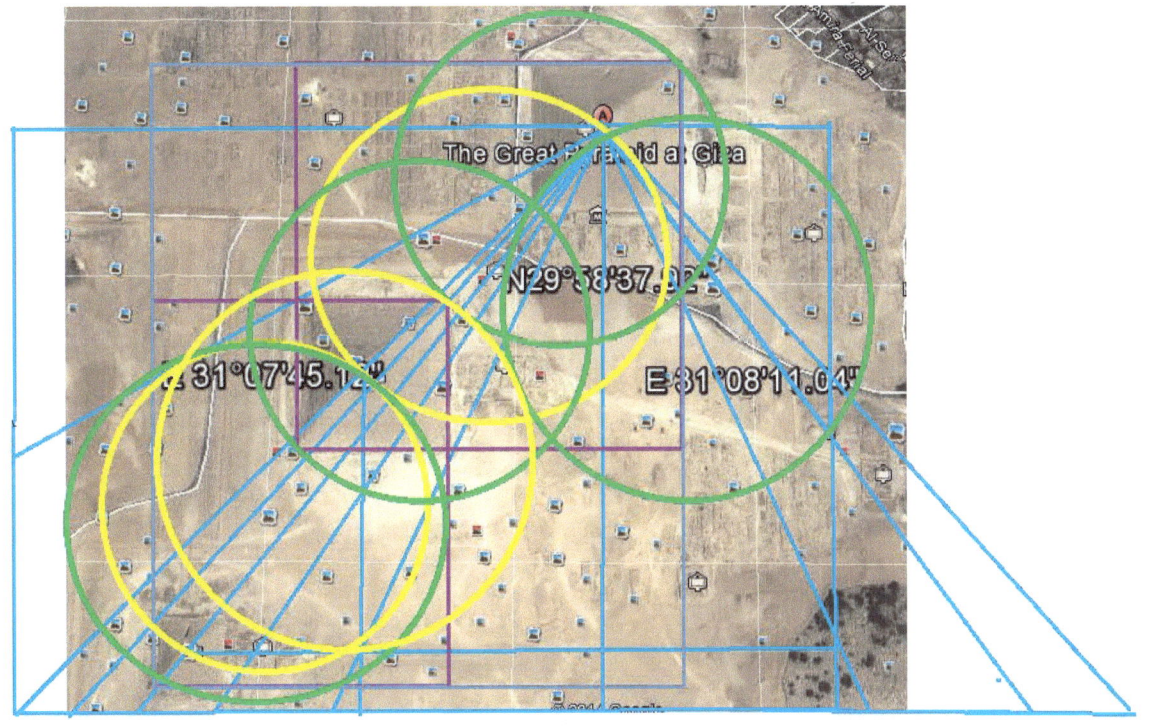

Figure A-22

T. Margulies, *Pyramid Geometry Design*

Law of cosines for sides for a spherical triangle,

$$\cos(a) = \cos(b)\cos(c) + \sin(b)\sin(c)\cos(A)$$

$$\cos(b) = \cos(c)\cos(a) + \sin(c)\sin(a)\cos(B)$$

$$\cos(c) = \cos(a)\cos b + \sin(a)\sin(b)\cos(C)$$

Law of cosines for the angles

$$\cos(A) = -\cos(B)\cos(C) + \sin(B)\sin(C)\cos(a)$$

$$\cos(B) = -\cos(C)\cos(A) + \sin(C)\sin(A)\cos(b)$$

$$\cos(C) = -\cos(A)\cos(B) + \sin(A)\sin(B)\cos(c)$$

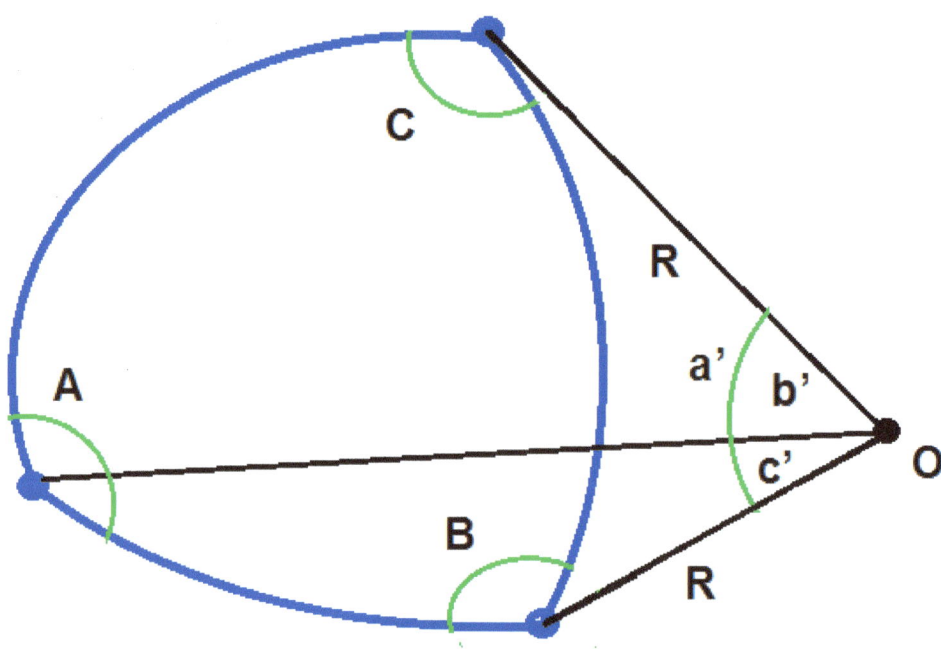

Figure A-23

Consider the surface of a sphere of radius R , centered at a point $O = (0,0,0)$, with a spherical triangle with vertices A, B , and C that also describe the angles. That is, A denotes the dihedral angle between the planes AOB and AOC; B is the angle between OC and AOB ; C is the angle between BOC and AOC.

The arc-lengths of the triangle sides are, $a = Ra'$, $b = Rb'$, and $c = Rc'$

T. Margulies, *Pyramid Geometry Design*

Angular lengths are given by ', b', and c' where $a' \equiv \angle BOC$, $b' \equiv \angle COA$, and $c' \equiv \angle AOB$ measured in radians.

The separation distance between the Stars of the Orion's belt were estimated by spherical trigonometry.

$$\text{Cos}(\gamma) = \cos\left(\frac{\pi}{2} - \delta_1\right)\cos\left(\frac{\pi}{2} - \delta_2\right) + \sin(\frac{\pi}{2} - \delta_1)\sin(\frac{\pi}{2} - \delta_2)\cos(RA_1 - RA_2)$$

$$\cos(\gamma) = \sin(\delta_1)\sin(\delta_2) + \cos(\delta_1)\cos(\delta_2)\cos(RA_1 - RA_2)$$

since $\cos\left(\frac{\pi}{2} - \alpha\right) = \sin(\alpha)$ and $\cos(-\alpha) = \cos(\alpha)$.

Right Ascension, RA, is measured positive in the eastward direction from the Vernal Equinox that represents the intersection of the ecliptic and equator planes. Declination is similar to latitude and is expressed in degrees, minutes, and seconds north or south (negative) of the equator. Conversions are used for the twenty-four hour day with hour-minute-seconds units and degrees-minutes seconds of a circle with angular measure of 360 degrees.

Table A-1

Star	RA	DEC	mag	RA diff	Cos(Sep)	Sep [deg]
Alnitak	85.20000000	-1.94277778	1.74000000	0.00000000		
Alnilam	84.05333333	-1.20191667	1.69000000	1.14666667	0.99971631	1.36481182
Minitaka	83.00166667	-0.29916667	2.25000000	1.05166667	0.99970747	1.38591041

Analytic Hierarchy Choices With Matrix Analysis

The analytic process presented by Thomas Saaty is illustrated in application to the three hypotheses for pyramid complex construction: Orion belt brightness, Cygnus belt brightness, and ancestral line. Each are given factors and weights in the calculation using Microsoft Excel. The valuations are close and show a preference for the Orion Constellation as a probable design motivation choice with Cygnus a close second.

Further the ranges may be adjusted to a scale range of zero to one by using the equation,

$$scale\ ranged\ score = \frac{NUB - NLB}{OUB - OLB}(original\ score - OLB) + NLB$$

T. Margulies, *Pyramid Geometry Design*

Table A-2

Preference Matrix							
Criteria ↓	Choice 1	Choice 2	Choice 3	OLB	OUB		
factor 1	7	7	0	0	7		
factor 2	2	2	9	0	9		
factor 3	3	2	1	1	3		
Sum	12	11	10	Total Sum=	33		
Norm. Score	31.82%	33.33%	34.85%		100.00%		
	36.36%	33.33%	30.30%				
NUB =	1	New upper bound		OLB:	Original lower bound		
NLB =	0	New lower bound		OUB:	Original upper bound		
Matrix M (3x3)		Re-Scaled Ranges					
Criteria ↓	Choice 1	Choice 2	Choice 3				
factor 1	1.000	1.000	0.000				
factor 2	0.222	0.222	1.000				
factor 3	1.000	0.500	0.000				
Sum	2.22	1.72	1.00	Total Sum =	4.944444		
Norm. Score	27.53%	32.58%	39.89%		100.00%		
Percent	44.94%	34.83%	20.22%		W^TM Alternatives		
Criteria ↓	Level Import.	Weight W	Wt. %	Criteria ↓	Choice 1	Choice 2	Choice 3
factor 1	7	0.538	53.85%	factor 1	0.538	0.538	0.000
factor 2	4	0.308	30.77%	factor 2	0.068	0.068	0.308
factor 3	2	0.154	15.38%	factor 3	0.154	0.077	0.000	Total
Sum	13	1.000	100.00%	Sum	0.761	0.684	0.308	1.752

W^T (1x3)				Percentage	43.41%	39.02%	17.56%
0.538	0.308	0.154					
	0.538	0	0				
	0	0.308	0				
	0	0	0.154				
S (1x3)							
	1	1	1				

T. Margulies, *Pyramid Geometry Design*

NUB new upper bound, NLB new lower bound, OUB old upper bound, OLB old lower bound

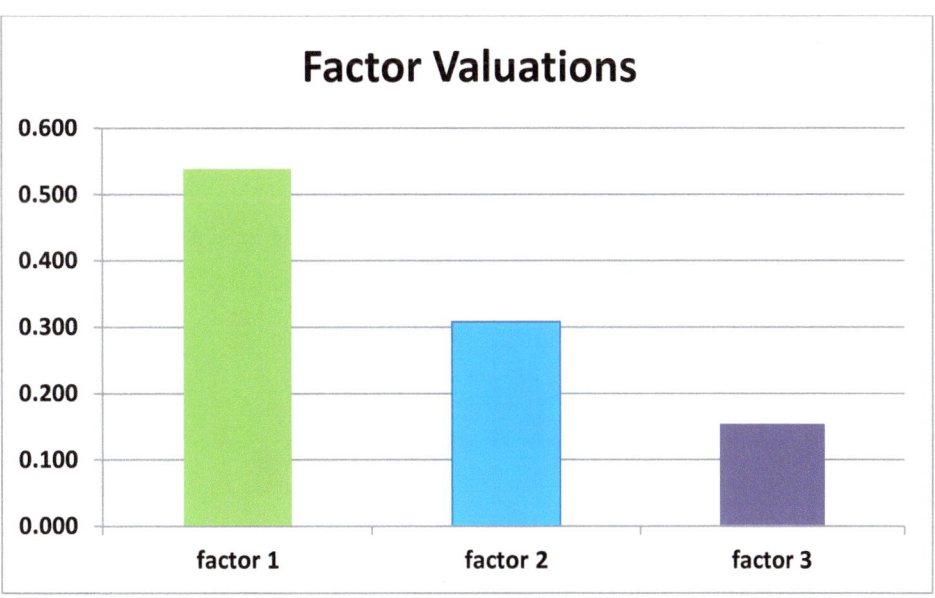

Figure A-24

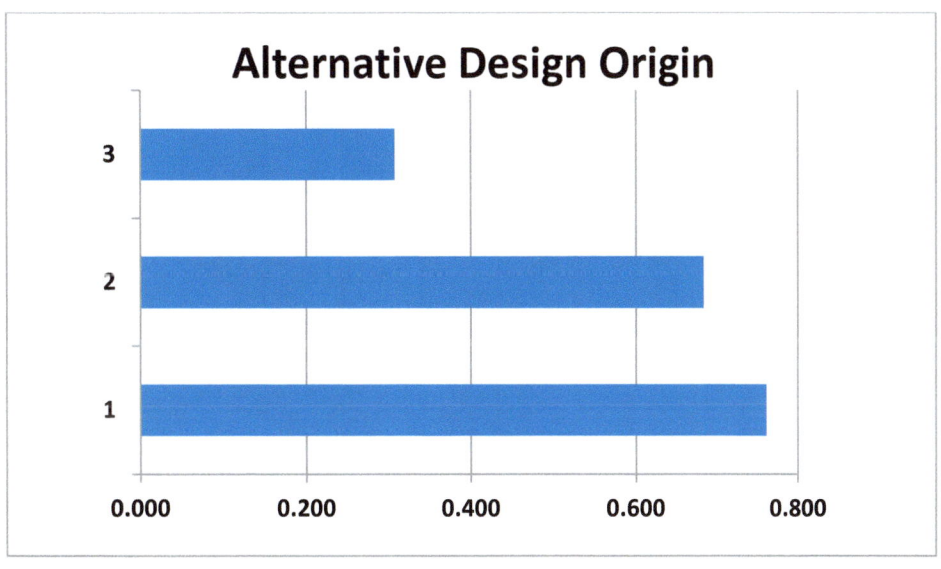

Figure A-25

Two Port Optics Matrix Transfer Models

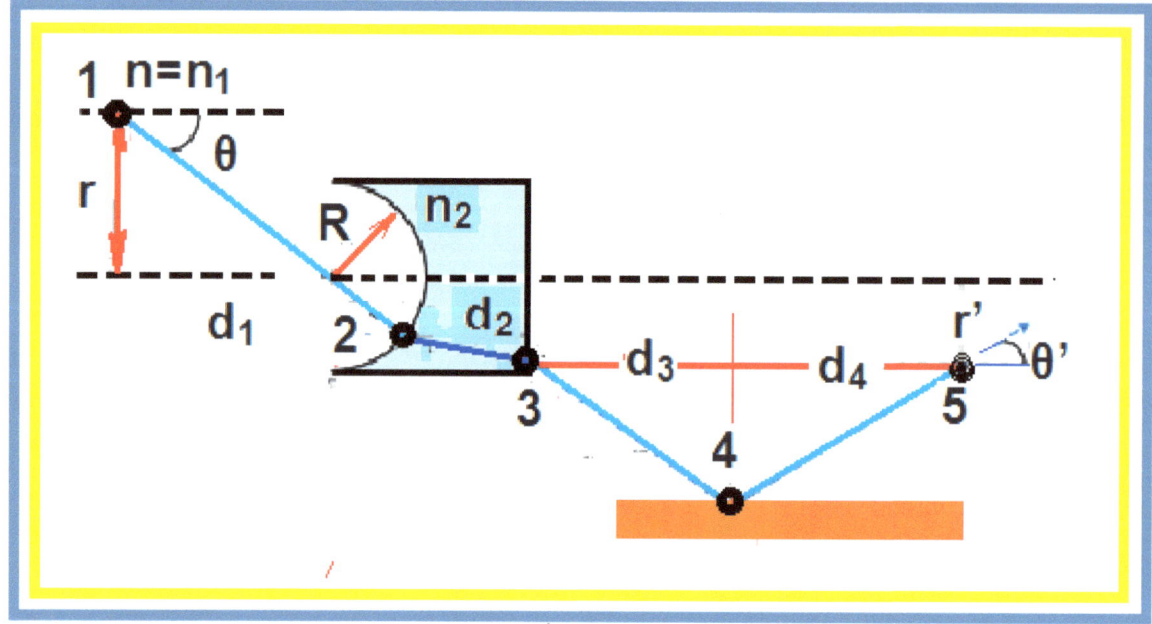

Figure A-26

$$\underline{\underline{M}} = \begin{pmatrix} 1 & d_4 \\ 0 & 1 \end{pmatrix} \begin{pmatrix} 1 & 0 \\ 0 & 1 \end{pmatrix} \begin{pmatrix} 1 & d_3 \\ 0 & 1 \end{pmatrix} \begin{pmatrix} 1 & 0 \\ 0 & \frac{n_2}{n_1} \end{pmatrix} \begin{pmatrix} 1 & d_2 \\ 0 & 1 \end{pmatrix} \begin{pmatrix} \frac{1}{n_1} & 0 \\ \frac{n_1 - n_2}{R \cdot n_2} & \frac{n_1}{n_2} \end{pmatrix} \begin{pmatrix} 1 & d_1 \\ 0 & 1 \end{pmatrix}$$

1. Transfer matrix of the medium between points 1 and 2 (with constant index of refraction n_1)
2. Transfer matrix of the curved interface at point 2
3. Transfer matrix of the medium between points 2 and 3 (with constant index of refraction n_2)
4. Transfer matrix of the flat interface at point 3
5. Transfer matrix of the medium between points 3 and 4 (with constant index of refraction n_1)
6. Transfer matrix of the flat mirror at point 4
7. Transfer matrix of the medium between points 4 and 5 (with constant index of refraction n_1)

Let $T' = tan\theta'$ and $T = tan\theta$

$$\begin{pmatrix} r' \\ T' \end{pmatrix} = \underline{\underline{M}} \begin{pmatrix} r \\ T \end{pmatrix}$$

The optical elements are cascaded with arrays of dimension 2x2 (2 rows by two columns) whose matrix multiplication models their transfer effects on the light including transmission, refraction, and reflection phenomena typically for small angle paraxial approximation $tan\theta \approx \theta$.

Table A-3

Transmission/Reflection/Refraction	Matrix Ray Model	Definitions
Transmission in free space or medium of constant refractive index	$\begin{pmatrix} 1 & d \\ 0 & 1 \end{pmatrix}$	d = distance
Refraction at a flat interface	$\begin{pmatrix} 1 & 0 \\ 0 & \dfrac{n_1}{n_2} \end{pmatrix}$	n_1= initial refractive index n_2= final refractive index
Refraction at a curved interface	$\begin{pmatrix} 1 & 0 \\ \dfrac{n_1 - n_2}{Rn_2} & \dfrac{n_1}{n_2} \end{pmatrix}$	R = radius of curvature
Reflection from a flat mirror	$\begin{pmatrix} 1 & 0 \\ 0 & 1 \end{pmatrix}$	2x2 Identity matrix
Reflection from a curved mirror	$\begin{pmatrix} 1 & 0 \\ \dfrac{-3}{R} & 1 \end{pmatrix}$	R radius of curvature, R >0 for convex
Thin lens	$\begin{pmatrix} 1 & 0 \\ \dfrac{-1}{f} & 1 \end{pmatrix} \begin{pmatrix} 1 & 0 \\ -\dfrac{1}{f} & 1 \end{pmatrix}$	f = focal length of lens where f > 0 for convex/positive (converging) lens where the focal length is much greater than the lens thickness

Matching Spot Luminosities as a signal on background reflectances as noise in a forced binary choice can be addressed within a statistical framework for psychoanalysis popularized by John Swets and Green. The human receiver's visual sensory system decides on the brightness in comparison. Illustrative data for calculation.

Central Force Inverse square Range Planetary Orbit Elliptical Motion

Consider two spherical celestial bodies with gravitation acting between their centers using polar coordinates (r, θ). The Lagrangian function $L = 0.5\mu(\dot{r}^2 + r^2\dot{\theta}^2) - V(r)$ where $\mu = \frac{m_1 m_2}{m_1 + m_2}$.

The equations of motion for the two variables r and θ are, respectively:

$$\frac{d}{dt}\left(\frac{\partial L}{\partial \dot{r}}\right) - \frac{\partial L}{\partial r} = \mu\ddot{r} - \mu r\dot{\theta}^2 + \frac{\partial V}{\partial r} = 0 \qquad (1)$$

$$\frac{d}{dt}\left(\frac{\partial L}{\partial \dot{\theta}}\right) - \frac{\partial L}{\partial \theta} = \frac{d}{dt}\left(\mu r^2\dot{\theta}\right) = 0 \qquad (2)$$

$$\mu r^2\dot{\theta} = constant = l \qquad (3)$$

$$\mu\ddot{r} - \frac{l^2}{\mu r^3} + \frac{\partial V}{\partial r} = 0 \quad \text{or} \quad \mu\ddot{r} + \frac{\partial P}{\partial r} = 0$$

T. Margulies, *Pyramid Geometry Design*

since $\dot{\theta} = \dfrac{d\theta}{dt} = \dfrac{l}{\mu r^2}$ and defining the new potential, $\quad P = \dfrac{l^2}{2\mu r^2} + V.$

Multiplying by $\dot{r}$ obtains

$$\mu\dot{r}\ddot{r} + \frac{\partial P}{\partial r}\dot{r} = 0 \quad \text{where} \quad \dot{r} = \frac{dr}{dt}$$

$$\frac{d}{dt}\left[\frac{\mu}{2}(\dot{r})^2 + P\right] = 0$$

or energy E is a constant, $\qquad \dfrac{\mu}{2}(\dot{r})^2 + P = E$

Solving for $\dot{r}$, $\dot{r} = \pm\sqrt{\dfrac{2}{\mu}(E - P)}$. Now using a change of variable differentiation, $\dfrac{d}{dt} = \dfrac{l}{\mu r^2}\dfrac{d}{d\theta}$

$\dfrac{l}{\mu r^2}\dfrac{dr}{d\theta} = \pm\sqrt{\dfrac{2}{\mu}(E - P)}$ so that separating, $\dfrac{l}{\mu\sqrt{\dfrac{2}{\mu}\left(E - \left[\frac{l^2}{2\mu r^2} + V\right]\right)}}\dfrac{dr}{r^2} = \pm d\theta$.

The central force potential decreases as the range, $V = -\dfrac{\kappa}{r}$. Define $\qquad w = \dfrac{1}{r} \qquad dw = -\dfrac{dr}{r^2}$

$\dfrac{l}{\mu\sqrt{\dfrac{2}{\mu}\left(E - \frac{l^2}{2\mu r^2} + \frac{\kappa}{r}\right)}}\dfrac{dr}{r^2} = \pm d\theta$ or rewriting,

$$\frac{-dw}{l\sqrt{-w^2 - \dfrac{2\kappa\mu}{l^2}w - \dfrac{2\mu E}{l^2}}} = \pm d\theta$$

Integrating,

$$-\left[\left(w + \frac{\kappa\mu}{l^2}\right)^2 - \left[\left(\frac{\kappa\mu}{l^2}\right)^2 - \frac{2\mu E}{l^2}\right]\right] = -[z^2 - A^2] \qquad z \equiv w + \frac{\kappa\mu}{l^2}, \quad dz = dw, \quad A^2 = \left[\left(\frac{\kappa\mu}{l^2}\right)^2 - \frac{2\mu E}{l^2}\right]$$

$\dfrac{-dz}{l\sqrt{A^2 - z^2}} = \pm d\theta = \dfrac{-dz}{lA\sqrt{1 - \frac{z^2}{A^2}}}$ Let $q \equiv \dfrac{z}{A}$, $\pm d\theta = \dfrac{-dq}{lA\sqrt{1-q^2}}$. Substitute $q = \sin(\phi)$, $dq = \cos(\phi)d\phi$

$\pm d\theta = \dfrac{-dq}{lA\sqrt{1-\sin^2(\phi)}} = \dfrac{-\cos\phi \cdot d\phi}{lA \cdot \cos\phi} = \dfrac{-d\phi}{lA}$. Then $\pm\theta\big]_0^1 = -\dfrac{\phi}{lA}\Big]_0^1 = $ or $\pm\theta\big]_0^1 = -\dfrac{1}{lA}\sin^{-1}\left(\dfrac{z}{A}\right)\Big]_0^1 = -\dfrac{1}{lA} \cdot$

$$\sin^{-1}\left(\frac{w + \frac{\kappa\mu}{l^2}}{\sqrt{\left(\frac{\kappa\mu}{l^2}\right)^2 - \frac{2\mu E}{l^2}}}\right)\Bigg]_0^1$$

For the negative signed case,

$$\sin[lA(\theta_0 - \theta_1)] = \left(\frac{\left(\frac{\kappa\mu}{l^2}\right) + \frac{1}{r}}{\sqrt{\left(\frac{\kappa\mu}{l^2}\right)^2 - \frac{2\mu E}{l^2}}}\right)\Bigg]_0^1 \qquad \eta = \sqrt{p^2 - \frac{2\mu E}{l^2}}, \qquad \eta \cdot \sin[lA(\theta_0 - \theta_1)] - \left(\frac{\kappa\mu}{l^2}\right) = \frac{1}{r}\Big]_0^1 = \frac{1}{r} - \frac{1}{r_0} = \frac{r_0 - r}{r r_0}$$

$$\left(\frac{\kappa\mu}{l^2}\right) = \quad \text{and} \quad \frac{r}{\left(1 - \frac{r}{r_0}\right)} = \frac{1}{\eta \cdot \sin[lA(\theta_0 - \theta_1)] - p}$$

T. Margulies, *Pyramid Geometry Design*

$$\frac{y}{(1-y)} = \frac{1}{\eta \cdot r_0 \cdot sin[lA(\theta_0 - \theta_1)] - p} \qquad \text{Define} \quad y = \frac{r}{r_0} \quad \text{and} \quad \frac{1}{\eta \cdot r_0 \cdot sin[lA(\theta_0 - \theta_1)] - p} = T$$

Then, $\quad \dfrac{y}{(1-y)} = T$

$$y = T(1-y) = T - Ty \quad y(1+T) = T \quad , \quad y = \frac{T}{(1+T)} = \frac{1}{1+\frac{1}{T}}, \quad \text{Finally,} \quad y = \frac{1}{1+\eta \cdot r_0 \cdot sin[lA(\theta_0 - \theta_1)] - p}$$

Table A-7

Planet	e	a semi-major axis AU	b	c
Mercury	0.206	0.387	0.264664	0.079722
Venus	0.007	0.723	0.719491	0.005061
Earth	0.017	1	0.991464	0.017
Mars	0.093	1.52	1.472766	0.14136
Jupiter	0.048	5.2	5.175944	0.2496
Saturn	0.056	9.54	9.511959	0.53424
Uranus	0.046	19.18	19.15699	0.88228
Neptune	0.01	30.07	30.065	0.3007
Pluto	0.248	39.44	39.3158	9.78112

The equation for elliptical paths with the Sun at one focus is given by,

$$\frac{(\zeta_1 - c_{Sun})^2}{a^2} + \frac{\zeta_2^2}{b^2} = 1$$

T. Margulies, *Pyramid Geometry Design*

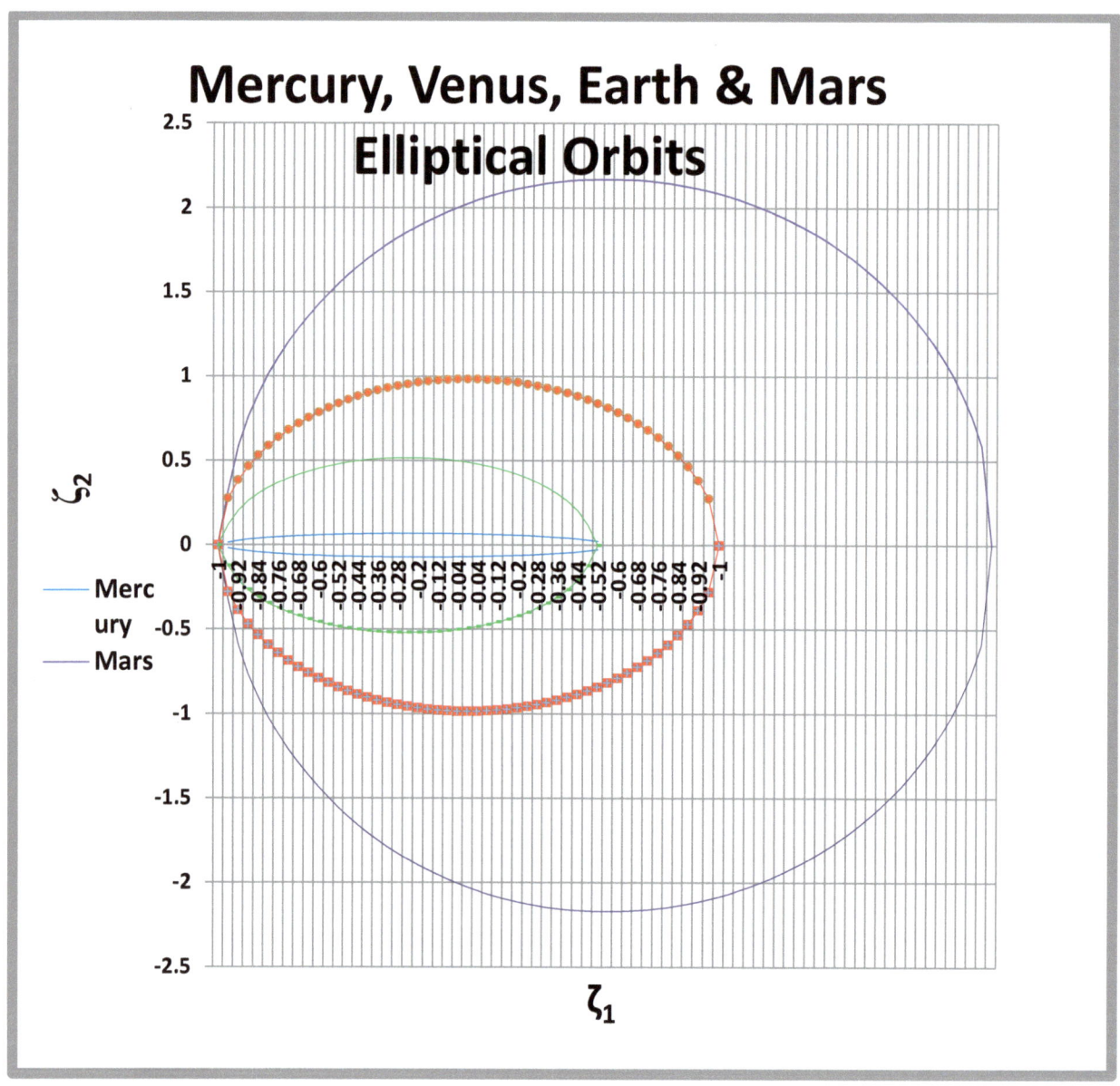

Figure A-27

On Kepler's Law of Planet Motion

The matrix equation is a system of linear equations such as for a polynomial function form with substituted (x,y) ordered pair data. For example, let $y = a_0 + a_1 x$ and substitute three observations,

$$\underline{\underline{A}} = \begin{pmatrix} (a_1)_1 & 1 \\ (a_1)_2 & 1 \\ (a_1)_3 & 1 \end{pmatrix}, \ \underline{x} = \begin{pmatrix} x_1 \\ x_2 \\ x_3 \end{pmatrix}, \ \underline{y} = \begin{pmatrix} y_1 \\ y_2 \\ y_3 \end{pmatrix}, \ \text{then} \ \ \underline{y} = \underline{\underline{A}}\underline{x}, \ \underline{\underline{A}}^T \underline{y} = \underline{\underline{A}}^T \underline{\underline{A}}\underline{x} \ , \ \underline{x} = \left[\underline{\underline{A}}^T \underline{\underline{A}}\right]^{-1} \underline{\underline{A}}^T \underline{y}$$

T. Margulies, *Pyramid Geometry Design*

The matrix approach is applied to Kepler's law that relates the square of the period of a planet's orbit to the cube of its mean radius from the Sun. Calculations used logarithmically transformed variables to fit a straight line whose slope corresponds to the exponent in the original untransformed equation.

Table A-8

Planet	Mercury	Venus	Earth	Mars	Jupiter	Saturn
distance	0.387	0.723	1	1.524	5.203	9.555
period	0.241	0.615	1	1.881	11.86	29.42

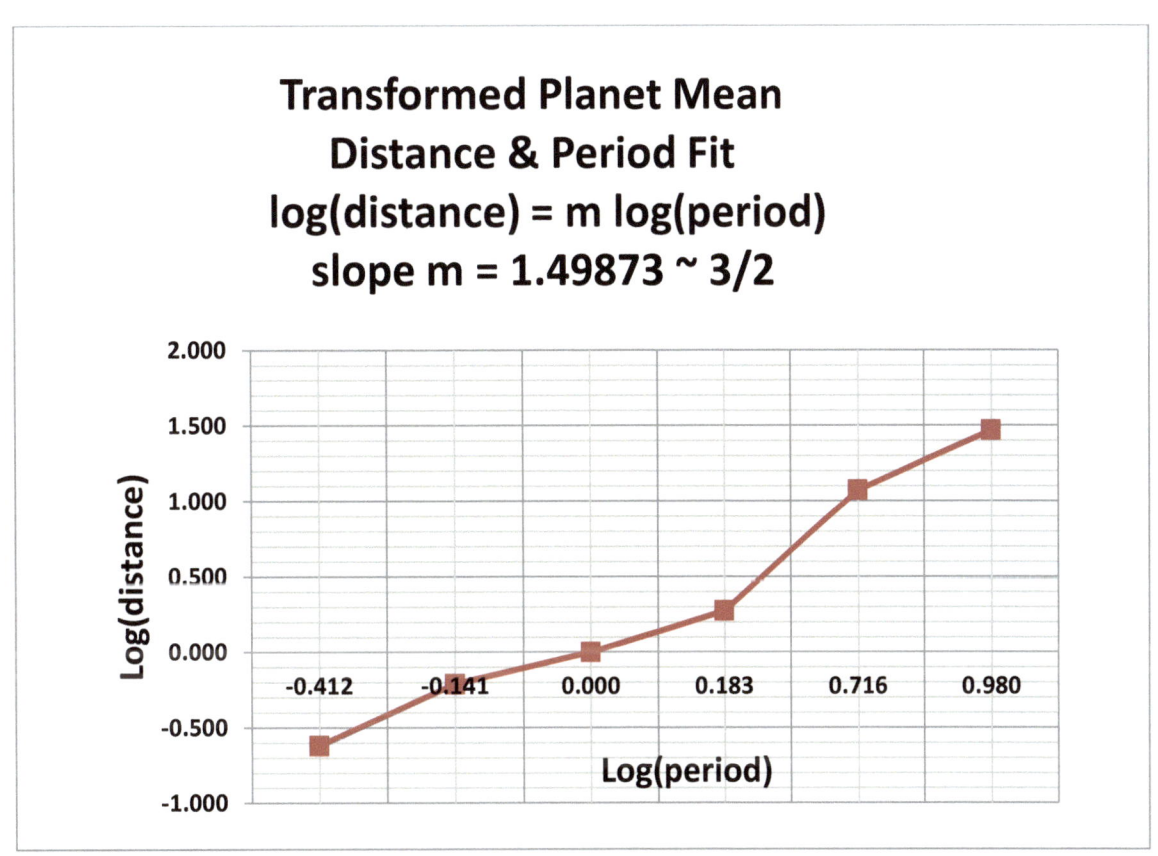

Figure A-28

T. Margulies, *Pyramid Geometry Design*

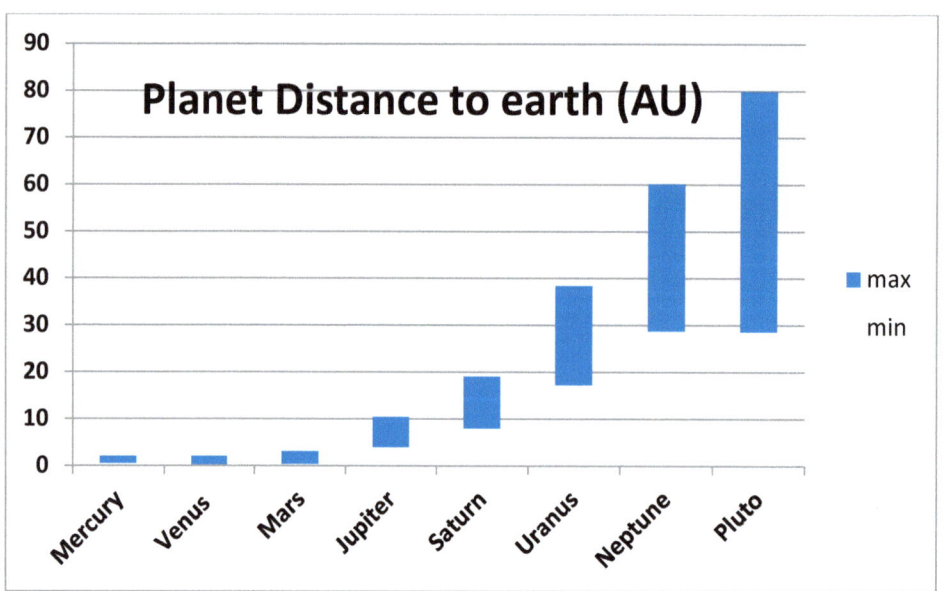

Figure A-29

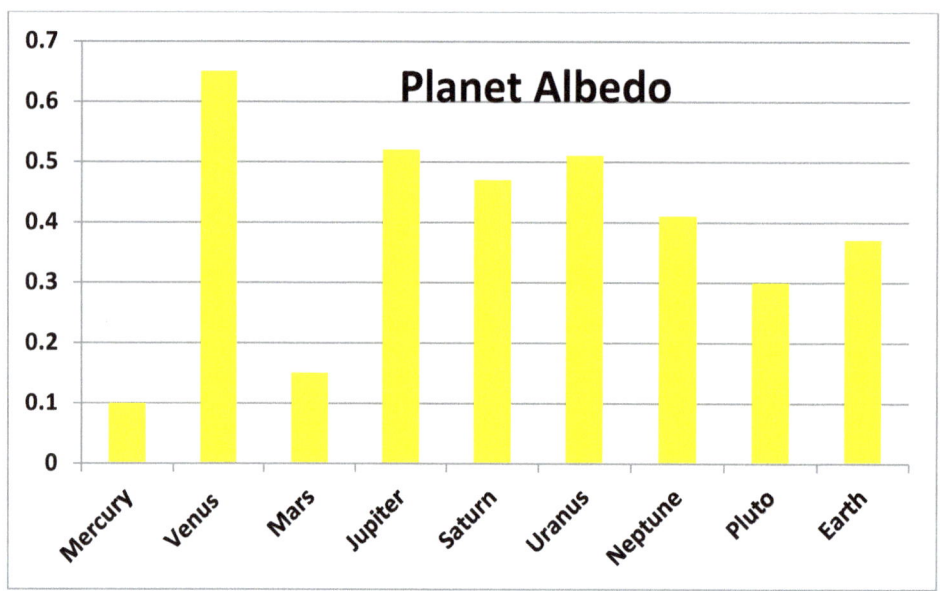

Figure A-30

The planet albedo is the visual geometric light reflectance of the planet expressed as the ratio of light reflected to that is all directions.

T. Margulies, *Pyramid Geometry Design*

Examples of Determinant Notation for Several Geometry and Algebra Problems

The equation for a line given by two points by their coordinate positions in a rectangular coordinate system.

$$\begin{vmatrix} x & y & 1 \\ x_1 & y_1 & 1 \\ x_2 & y_2 & 1 \end{vmatrix} = 0$$

The equation for a plane as determined by three points expressed as coordinates or ordered pairs (x,y).

$$\begin{vmatrix} x & y & z & 1 \\ x_1 & y_1 & z_1 & 1 \\ x_2 & y_2 & z_2 & 1 \\ x_3 & y_3 & z_3 & 1 \end{vmatrix} = 0$$

A conic section such as an ellipse, hyperbola, or parabola is analytically described by the equation,

$$, ax^2 + bxy + cy^2 + dx + ey + f$$

and may be fit by five coordinates to find the coefficients by,.

$$\begin{vmatrix} x^2 & xy & y^2 & x & y & 1 \\ x_1^2 & x_1 y_1 & y_1^2 & x_1 & y_1 & 1 \\ x_2^2 & x_2 y_2 & y_2^2 & x_2 & y_2 & 1 \\ x_3^2 & x_3 y_3 & y_3^2 & x_3 & y_3 & 1 \\ x_4^2 & x_4 y_4 & y_4^2 & x_4 & y_4 & 1 \\ x_5^2 & x_5 y_5 & y_5^2 & x_5 & y_5 & 1 \end{vmatrix} = 0$$

Area of a triangle described by its three vertices in coordinates.

$$A = \pm \frac{1}{2} \begin{vmatrix} x_1 & y_1 & 1 \\ x_2 & y_2 & 1 \\ x_3 & y_3 & 1 \end{vmatrix}$$

Volume of a pyramid given by four vertices' coordinates

$$V = \pm \frac{1}{6} \begin{vmatrix} x_1 & y_1 & z_1 & 1 \\ x_2 & y_2 & z_2 & 1 \\ x_3 & y_3 & z_3 & 1 \\ x_4 & y_4 & z_4 & 1 \end{vmatrix}$$

T. Margulies, *Pyramid Geometry Design*

Special Relativity Principle for Light Waves

The special theory of relativity from physics by Albert Einstein in hisearly writings captured many minds on the subject with the use of algebra and his understanding of physics and measurement invariance with regard to frame. The word "algebra" comes from Arabic, "Al" meaning "the" and "gebar" meaning "to set" or "restitute". The word "algebrista" in Spain refers to a bone setter. The use of the modifier special refers to a generalization, however for frames in uniform translation which was reformulated to account for further variations, in particular with mathematics of differential geometry and tensor analysis.

Consider the distance traveled in the two frames of reference

$x' = c \cdot t'$ and $x = c \cdot t$ or re-write as

$$x' - c \cdot t' = 0 \; x - c \cdot t = 0$$

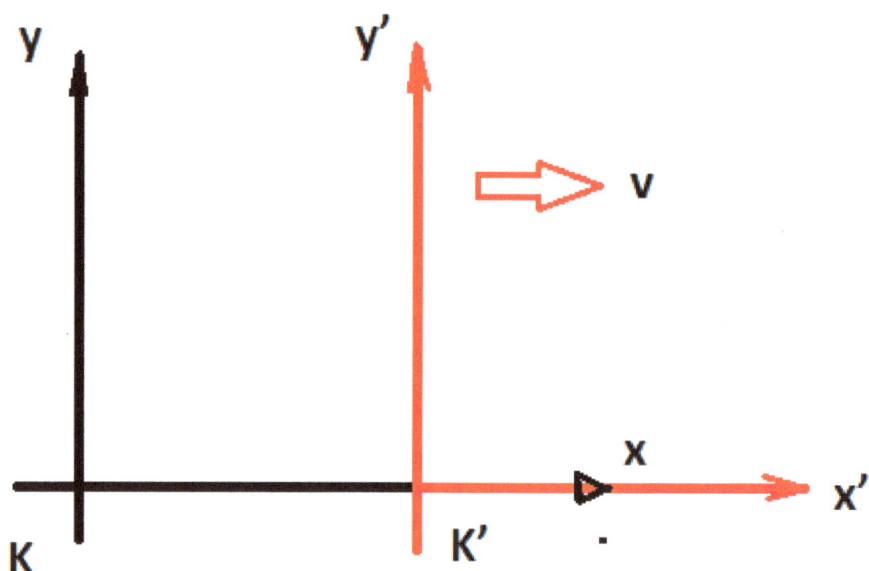

Figure A-31

Distances are denoted by x, x' and chronological clock times by

t, t'. Assuming one is a constant multiple of the other, using

constants λ, μ as in parallel $x' - c \cdot t' = \lambda(x - c \cdot t)$

$$x' + c \cdot t' = \mu(x + c \cdot t)$$

T. Margulies, *Pyramid Geometry Design*

Add and subtract the system of equations obtains,

$$2x' = (\lambda + \mu)x - c \cdot t(\lambda - \mu)$$

$$-2c \cdot t' = (\lambda - \mu)x - c \cdot t(\lambda + \mu)$$

Let one define $\quad a \equiv 0.5(\lambda + \mu),\quad$ and $\quad b \equiv 0.5(\lambda - \mu)$

Then the system of equations become, $\quad x' = a \cdot x - b \cdot c \cdot t\quad$ and

$$x = \frac{a \cdot c \cdot t}{b} - \frac{c \cdot t'}{b}$$

Applying a condition $\quad x' = 0,\quad$ yields $x = \dfrac{b \cdot c}{a} \cdot t\quad$ which can be used to define the relative velocity of

the origins of the two frames. That is, $\quad x = v \cdot t\quad$ where $v = \dfrac{b \cdot c}{a}$

so that $\qquad \dfrac{v^2}{c^2} = \dfrac{b^2}{a^2}$

Performing further algebra manipulation using the properties of real numbers and rules of exponents follows.

$$x' = a \cdot x - b \cdot c \cdot \left(\frac{b \cdot x}{a \cdot c}\right) \quad x' = \left(a - \frac{b^2}{a}\right) \cdot x,$$

$$x' = \left(\frac{a^2 - b^2}{a}\right) \cdot x, \quad x' = \left(\frac{1 - \dfrac{b^2}{a^2}}{\dfrac{1}{a}}\right) \cdot x \quad x' = a \cdot \left(1 - \frac{b^2}{a^2}\right) \cdot x, \quad x' = a \cdot \left(1 - \frac{v^2}{c^2}\right) \cdot x$$

Intervals of lengths among the two frames are related by

$$\Delta x' = a \cdot \left(1 - \frac{v^2}{c^2}\right) \cdot \Delta x \quad x' = \frac{a}{x}$$

T. Margulies, *Pyramid Geometry Design*

This when combined with Einstein's thinking on light travels and its equivalence of measurement, or invariance regardless of the frame used obtains an equation for the constant a.

In particular, $\Delta x' = \Delta x$ obtains,

$$\frac{x}{a} = a \cdot \left(1 - \frac{v^2}{c^2}\right) \cdot x \quad a^2 = \left(1 - \frac{v^2}{c^2}\right)^{-1}$$

or , so that

$$a = \left(1 - \frac{v^2}{c^2}\right)^{-0.5}$$

Otherwise

independent measurements of v and c would obtain estimates of a. Substituting the value obtains,

$$\Delta x' = \left(1 - \frac{v^2}{c^2}\right)^{-0.5} \cdot \left(1 - \frac{v^2}{c^2}\right) \cdot \Delta x \quad \Delta x' = \left(1 - \frac{v^2}{c^2}\right)^{0.5} \cdot \Delta x$$

Table A-9

v/c	L/L₀	t/t₀	λ	Color
0	1	1	450.00	"Blue"
0.25	0.9682	1.0328	562.50	"Yellow"
0.5	0.8385	1.1926	843.75	near infrared
0.75	0.5546	1.8030	1476.56	infrared
0.999	0.0248	40.3263	2951.65	infrared

The expanding universe and Doppler effects for relative motions alter the luminosity functions.

The color spectrum observing sunlight with a grating spectrometer yielded lines that were comparable to that produced by terrestrial elements in the 1860's.

The spectrum of thermal radiation for wavelength, λ, from a blackbody at a temperature T (Kelvin) is quantitatively presented in Planck's law,

$$R(\lambda, T) = \frac{2\pi hc^2}{\lambda^5 \left(e^{\frac{hc}{\lambda kT}} - 1\right)}$$

T. Margulies, *Pyramid Geometry Design*

Integrating the radiance over all wavelengths obtains the total radiated power as the Stefan-Boltzmann's law, $P_{rad} = \sigma A T^4$ where A denotes surface area and σ is a constant.

Further insight into radiation color derives from finding the wavelength that maximizes the radiation referred to as Wein's radiation law, $\lambda_{max} = T = 2.898 \ 10^{-3} \ [mK]$. This captures the observations of color from an incandescent object, spanning from red through orange and yellow to white hot as temperature increases.

Let the proper separation distance, $L(t)$, between two galaxies be expressed by a universal expansion factor, $a(t)$ by, $L(t) = L_0 \cdot a(t)$ or $L_0 = \frac{L(t)}{a(t)}$ with constant L_0. Taking the time differential obtains, $\dot{L}(t) = L_0 \cdot \dot{a}(t) = L\frac{\dot{a}(t)}{a}$. The ratio defines a Hubble variable,

$$H = \frac{\dot{a}(t)}{a}.$$

Observations of color shifts in the wavelengths of electromagnetic signals provide evidence of this red-shift or expansion of the universe. The red-shift parameter, , ratios the observed to emitted wavelengths and to first order is a Doppler shift. $z = \frac{\lambda_O}{\lambda_E} - 1$. Einstein's experimental validation of general relativity relied on the perihelion precession of Mercury, the deflection of light by the Sun, and the red-shift of light.

Modeling Expansion-Contraction of the Universe

Hubble's observation of an expanding universe is revisited with a time-dependent metric for space-time in general relativity. Three equations for the perfect, non-dissipative fluid are described by Eqs. 2-4,

$$H^2 = g_s\rho - \frac{kc^2}{a^2} \quad \text{where} \quad g_s = \frac{8\pi G}{3}, \quad H = \frac{\dot{a}}{a} \tag{1}$$

$$\dot{\rho} + 3H\left(\rho + \frac{p}{c^2}\right) = 0 \tag{2}$$

$$\frac{\ddot{a}}{a} - 0.5g_s\left(\rho + \frac{p}{c^2}\right) = 0 \tag{3}$$

$$f(a) = \rho + \frac{p}{c^2}$$

$$\ddot{a} - 0.5g_s a\dot{a}f(a) = 0 \tag{4}$$

$$\frac{d}{dt}(\dot{a})^2 - g_s\frac{d}{dt}(a)^2 f(a) = 0 \tag{5}$$

$$(\dot{a})^2 - [g_s a^2 f(a)] + \int a^2 df = g(r) \tag{6}$$

$$df = \frac{\partial f}{\partial a}da$$

$$(\dot{a})^2 - [g_s a^2 f(a)] + \int a^2 \frac{\partial f}{\partial a}da = g(r) \tag{7}$$

T. Margulies, *Pyramid Geometry Design*

$$(\dot{a})^2 - h(a) = g(r)$$

$$h(a) = [g_s a^2 f(a)] + \int a^2 \frac{\partial f}{\partial a} da$$

$$\text{For } g(r) = 0, \quad \dot{a} = \sqrt{h(a)} \tag{8}$$

$$\int \frac{da}{\sqrt{h(a)}} = \int dt = t - t_0 \tag{9}$$

For illustrative analysis, let $f(a) = \beta a^{-3}$ so that $\frac{\partial f}{\partial a} = -3\beta a^{-4}$, and $h(a) = [g_s \beta a^{-1}] - 3\beta \int a^{-2} da$.

$$h(a) = [g_s \beta a^{-1}] + 3\beta a^{-1} + \gamma = \frac{\delta}{a} + \gamma, \text{ where } \delta \equiv g_s \beta + 3\beta$$

$$\int \frac{da}{\sqrt{\frac{\delta}{a} + \gamma}} = \int \sqrt{\frac{a}{\gamma a + \delta}} \, da = t - t_0 \tag{10}$$

$$\tag{11}$$

$$\frac{1}{\sqrt{\gamma}} \int \sqrt{\frac{u^2}{u^2 + \delta}} \, da = \frac{2}{\gamma\sqrt{\gamma}} \int \frac{u^2}{\sqrt{u^2 + \delta}} \, du \text{, with } u^2 = \gamma a, \ (2u)du = (\gamma) \, da \text{ ,}$$

$$\frac{2}{\gamma\sqrt{\delta\gamma}} \int \frac{u^2}{\sqrt{\left(\frac{u}{\delta}\right)^2 + 1}} \, du \text{ , } \frac{u}{\delta} = z \quad \frac{2\delta^3}{\gamma\sqrt{\delta\gamma}} \int \frac{z^2}{\sqrt{z^2 + 1}} \, dz \text{ , } \frac{u}{\delta} = z$$

$$\frac{2\delta^3}{\gamma\sqrt{\delta\gamma}} \int \frac{\tan(\theta)^2 \sec(\theta)^2}{\sqrt{\tan(\theta)^2 + 1}} \, d\theta \quad z = \tan(\theta), \ \frac{2\delta^3}{\gamma\sqrt{\delta\gamma}} \int \tan(\theta)^2 \sec(\theta) \, d\theta, \ \frac{2\delta^3}{\gamma\sqrt{\delta\gamma}} \int (\sec(\theta)^2 - 1)\sec(\theta) \, d\theta$$

$$\frac{2\delta^3}{\gamma\sqrt{\delta\gamma}} \left\{ \int \sec(\theta)^3 d\theta - \int \sec(\theta) \, d\theta \right\} =$$

$$\frac{2\delta^3}{\gamma\sqrt{\delta\gamma}} \{0.5\sec(\theta)\tan(\theta) + 1.5\ln[\sec(\theta) + \tan(\theta)]\} + cst$$

since, $\int \sec(\theta)^3 d\theta = 0.5\sec(\theta)\tan(\theta) + 0.5\ln[\sec(\theta) + \tan(\theta)] + cst$

$$\int \sec(\theta) d\theta = \ln[\sec(\theta) + \tan(\theta)] + cst$$

$$z = \tan(\theta) = \frac{u}{\delta} \text{ , } \theta = \tan^{-1}\left(\frac{u}{\delta}\right), \ u = \sqrt{\gamma a}, \ \theta = \tan^{-1}\left(\frac{\sqrt{\gamma a}}{\delta}\right)$$

T. Margulies, *Pyramid Geometry Design*

$$\frac{2\delta^3}{\gamma\sqrt{\delta\gamma}}\left\{0.5sec\left(tan^{-1}\left(\frac{\sqrt{\gamma a}}{\delta}\right)\right)\frac{\sqrt{\gamma a}}{\delta}+1.5ln\left[sec\left(tan^{-1}\left(\frac{\sqrt{\gamma a}}{\delta}\right)\right)+\left(\frac{\sqrt{\gamma a}}{\delta}\right)\right]\right\}+cst=t-t_0$$

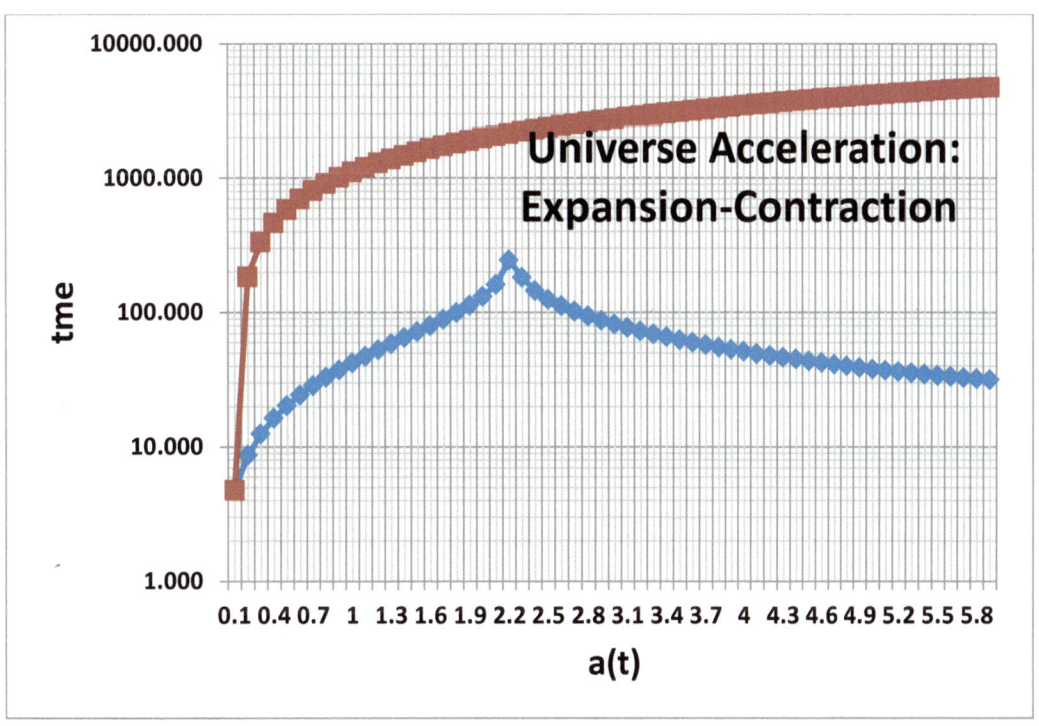

Figure A-32

Energy flux, F, at a given luminosity, angular size distance, and motion Doppler redshift is a function of frequency. For a uniform spherical bright source of radius r with the values integrated over all frequencies the energy flux varies as $(1+z)^{-4}$; that is, $F=\int_0^1 2\pi I\cdot\cos(\theta)\,d(\cos(\theta))=\pi I$.

Furthermore, The luminosity is the multiplicative product of the surface area of the sphere and F.

$$L=(4\pi r^2)\pi I=4\pi^2 r^2 I$$

The universe expands making the observed surface brightness given by $I_0=\frac{I}{(1+z)^4}$ and the angular radius of the source at the observer is $\theta=\frac{r(1=z)}{a_0\cdot r(z)}$. Then the observed energy flux is the product of the surface brightness and solid angle, $F=\pi\theta^2 I_0$ and $F_0=\frac{L}{4\pi(a_0\cdot r)^2(1+z)^2}$. As the source radius increases which is in the denominator the energy flux decreases.

A quantum mechanics relationship using wavelengths which are the Golden ratio has been documented. That is, the time-independent Schrodinger equation for a particle in one dimension is written,

T. Margulies, *Pyramid Geometry Design*

$$\frac{-\tilde{h}^2}{2m}\frac{d^2\psi}{dx^2} + V(x)\psi(x) = E\psi(x)$$ where λ is the wavelength, $\tilde{h} = \frac{h}{2\pi}$ is the Plank constant,

and c is the light speed. The energy $E = \frac{hc}{\lambda}$ is only a function of the space dimension x. The wave

function solutions are

$$\Psi(x,t) = \psi(x)e^{2\pi ict/\lambda}|\Psi(x,t)| = |\psi(x)|^2$$

With a phi relationship between wavelengths, $\lambda_{n+1} = \Phi\lambda_{n+2}$ one obtains,

$$\Psi_n(x,t) = \psi_n(x)e^{-2\pi ict/\Phi^2\lambda_{n+2}}$$

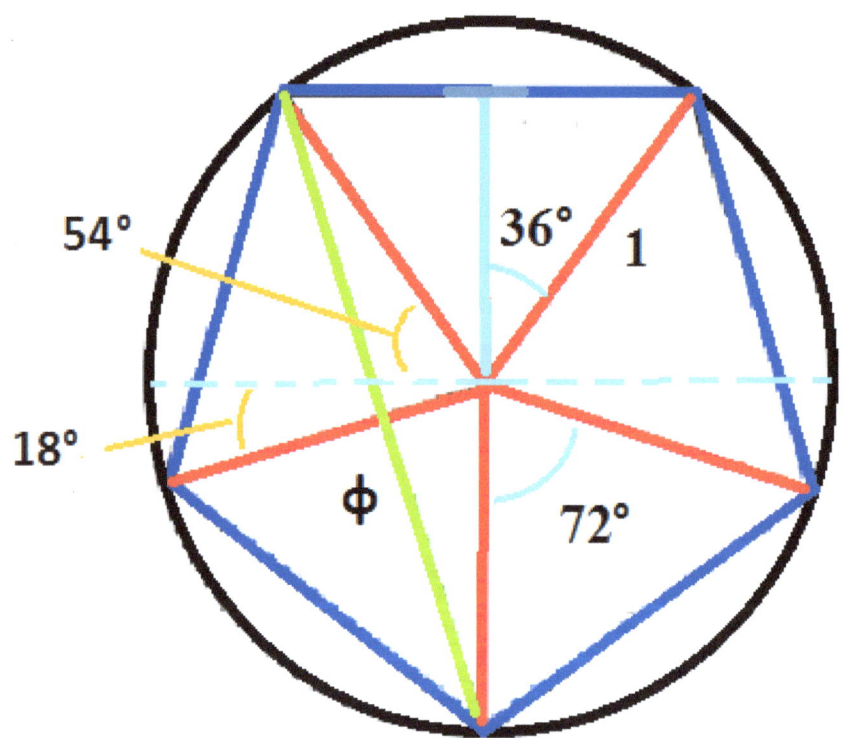

Figure A-33

$$\sin(18°) = \frac{\phi}{2}, \quad \sin(36°) = \frac{2}{\phi}, \quad \sin(54°) = \frac{\phi}{2}, \quad \sin(72°) = 2\phi$$

T. Margulies, *Pyramid Geometry Design*

A five sided polygon, or pentagon can be drawn with a diagonal of length φ. Further analysis of polygon math reveals interesting relationships using trigonometry.

An inscribed pentagon has five isosceles triangles with 360/5 = 72 degree vertical angles. The opposite side-length can be represented by $2 \cdot \sin 36 = \sqrt{4 - \phi^2}$.

Using the Pythagorean Theorem for an n-sided polygon of side-lengths one, and k as sketched in the diagram,

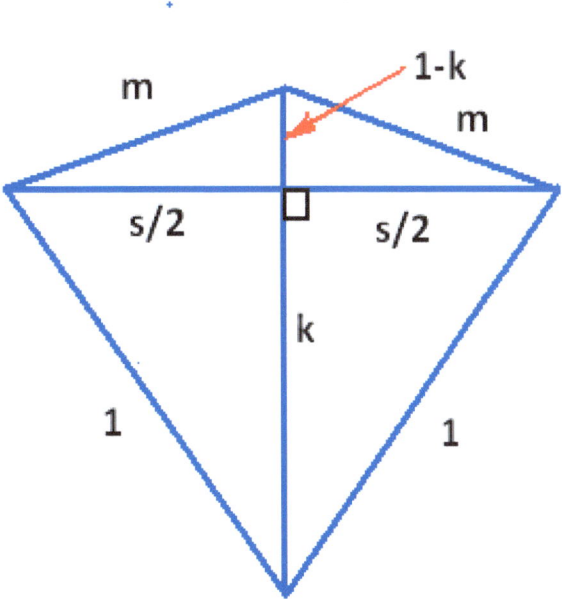

Figure A-34

$$1^2 = k^2 + \frac{s^2}{4} \ , \sqrt{1 - \frac{s^2}{4}} = k = \frac{\sqrt{4 - s^2}}{2}$$

$$1 - k = 1 - \frac{\sqrt{4 - s^2}}{2}$$

$$m^2 = (1 - k)^2 + \frac{s^2}{4},$$

$$m^2 = \left(1 - \frac{\sqrt{4 - s^2}}{2}\right)^2 + \frac{s^2}{4} = 1 - \sqrt{4 - s^2} + \frac{s^2}{4} + \frac{4 - s^2}{4} = 2 - \sqrt{4 - s^2}, \ m = \sqrt{2 - \sqrt{4 - s^2}}$$

$$m_{n\,poly} = \sqrt{2 - \sqrt{4 - s_n^2}}$$

This formula can be applied to estimate π.

n	$\pi \sim 3.14159265358979$
7	3.1415800371154200
8	3.1415894994663300
9	3.1415918650251800
10	3.1415924564182300
11	3.1415926044112300
12	3.1415926414673800
13	3.1415926488786100
14	3.1415926785235200
15	3.1415927971031900

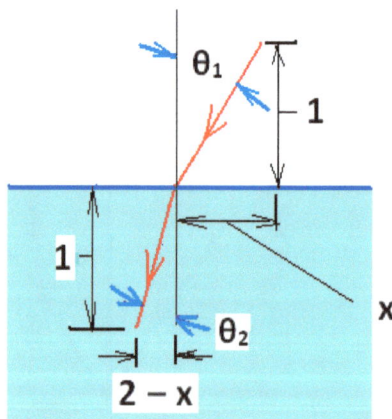

Figure A-35

Snell's Law

This derivation calculates the extrema (minimum) of the travel time T of the light rays traversing the two media. The media are homogeneous with constant speeds of traverse. Angles are defined relative to the normals to the interface by v_1 and v_2.

$$T(x) = \frac{\sqrt{1 + x^2}}{v_1} + \frac{\sqrt{1 + (2 - x)^2}}{v_2} \quad , \quad \text{Setting the first differential to zero and solving,} \quad T'(x) = 0$$

T. Margulies, *Pyramid Geometry Design*

$$T(x) = \frac{1}{v_1}\left(1 + x^2\right)^{0.5} + \frac{1}{v_2}\left(1 + (2-x)^2\right)^{0.5}$$ with using the chain and power law rules of

differentiation,

$$T'(x) = \frac{0.5}{v_1}\left(1 + x^2\right)^{-0.5}(2x) + \frac{0.5}{v_2}\left(1 + (2-x)^2\right)^{-0.5} 2(2-x)(-1)$$

Then

$$\frac{1}{v_1}\left(1 + x^2\right)^{-0.5}(x) = \frac{1}{v_2}\left(1 + (2-x)^2\right)^{-0.5}(2-x)$$

and

$$\frac{\left(1 + x^2\right)^{-0.5}(x)}{\left(1 + (2-x)^2\right)^{-0.5}(2-x)} = \frac{v_1}{v_2} = \frac{\left(1 + (2-x)^2\right)^{0.5} \cdot x}{\left(1 + x^2\right)^{0.5}(2-x)}$$

$$2 - x = \tan\theta_2 \text{ and } x = \tan\theta_1$$

$$\frac{v_1}{v_2} = \frac{\left(1 + \tan\theta_2{}^2\right)^{0.5} \cdot \tan\theta_1}{\left(1 + \tan\theta_1{}^2\right)^{0.5} \tan\theta_2},$$

$$\frac{v_1}{v_2} = \frac{\left(\sec\theta_2{}^2\right)^{0.5} \cdot \tan\theta_1}{\left(\sec\theta_1{}^2\right)^{0.5} \tan\theta_2}$$

Using the trigonometric identity, $\tan\theta^2 + 1 = \sec\theta^2$

$$\frac{v_1}{v_2} = \frac{\sec\theta_2 \cdot \tan\theta_1}{\sec\theta_1 \tan\theta_2} \quad \frac{v_1}{v_2} = \frac{\dfrac{1}{\cos\theta_2}\dfrac{\sin\theta_1}{\cos\theta_1}}{\dfrac{1}{\cos\theta_1}\dfrac{\sin\theta_2}{\cos\theta_2}} = \frac{\sin\theta_1}{\sin\theta_2} \quad \text{That is again,} \quad \frac{v_1}{v_2} = \frac{\sin\theta_1}{\sin\theta_2}$$

T. Margulies, *Pyramid Geometry Design*

R.B. Minton & R.T. Smith, *Calculus with early Transcendental Functions*, Third edition, McGraw Hill.

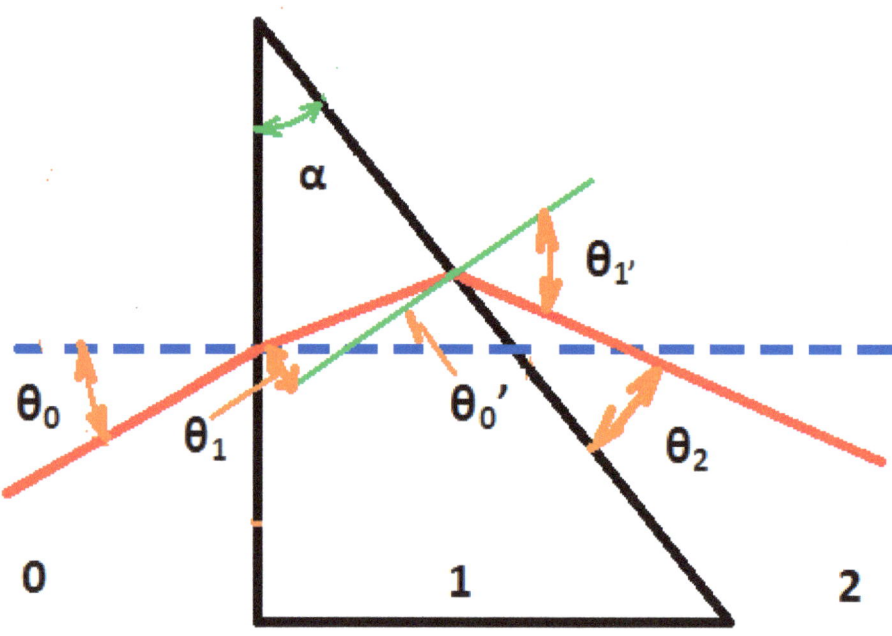

Figure A-36

A ray trace through a prism's regions 0, 1, and 2 have indices of refraction n_0, n_1, and n_2, and primed angles θ' to indicate the ray angles after refraction. Use Snell's law at each interface. For the prism shown, the indicated angles are given by $\quad n_1 sin\theta'_0 = n_0 sin\theta_0$, $\theta_1 = \alpha - \theta'_0$,

$n_2 sin\theta'_1 = n_1 sin\theta_1$, $\theta_2 = \theta'_1 - \alpha$, For air $n_0 = n_2 \approx 1$. Simplify by letting $n = n_1$, and define spread as $\quad \delta = \theta_0 + \theta_2$

$$= \theta_0 + arcsin\left(n \cdot sin\left(\alpha - arcsin\left(\frac{sin\theta_0}{n}\right)\right)\right) - \alpha \quad \text{Using the small angle approximations}$$

$sin\theta \approx \theta, = \beta$.

$$\delta = \theta_0 - \alpha + \left(n\left(\alpha - \frac{\theta_0}{n}\right)\right) = \theta_0 - \alpha + n \cdot \alpha - \theta_0 = (n-1)\alpha$$

Cartesian coordinates calculated from the spherical coordinates (*radius r, inclination* θ, *azimuth* φ), are:

$x = r \cdot sin\theta \cdot cos\varphi, \; y = r \cdot sin\theta \cdot sin\varphi , \; z = r \cdot cos\theta$

T. Margulies, *Pyramid Geometry Design*

Atmospheric transmission across layers with differing indices of refraction steer the direction of light. Assuming homogeneous constant temperature, pressure, and humidity a simple estimate is made by Snell's law across a layer of a single index of refraction. Let

$\rho = z_t - z_0$ where z_t denotes the true zenith angle and z_0 the observed zenith angle of the star

The angle of refraction for small angle, z_0,

$$\rho = (\mu - 1) \cdot tan(z_0) \ [radians] \qquad or \qquad \rho = 206265(\mu - 1) \cdot tan(z_0) \ [arcseconds]$$

More generally, $\quad \rho = sin^{-1}(\mu \cdot sin(z_0)) - z_0.$

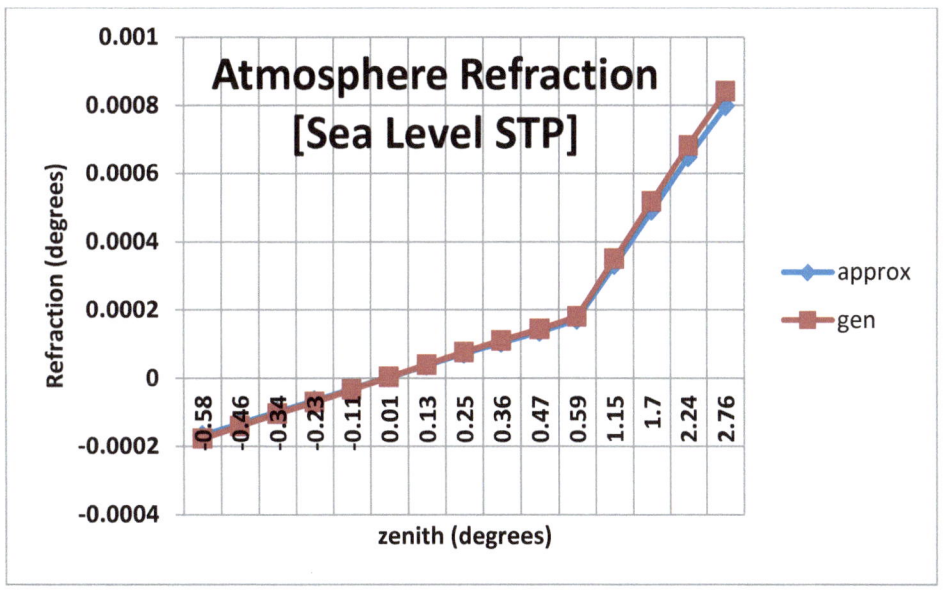

Figure A-37

Parallax from distance and rotation, as well as, aberrations contribute to the observed positions,

Fractional Calculus Kinetics Model

A star physics model for the high-temperature luminous surface is kinetics based. First the first order differential equation is presented followed by a generalization to fractional calculus.

$\frac{dn}{dt} = -c \cdot n(t)$ where $c > 0$ accounts for the probable production and depletion of the number density of each species and $n(t = 0) = n_0$.

Then, $\qquad\qquad\qquad\qquad \frac{dn}{n} = -c \cdot dt \ $ and $ \ ln(n) - ln(n_0) = -ct$

Rewriting and exponentiating, $\quad ln\left(\frac{n}{n_0}\right) = -ct$, and $\frac{n}{n_0} = e^{-ct}.$

T. Margulies, *Pyramid Geometry Design*

Since by integration, $dn = -c \cdot n(t)dt,$

$$-c \cdot n(t) = {_0\partial_t^{-1}} n(t)$$

A Riemann-Louisville fractional calculus definition proceeds as follows. Let Re $v > 0$, f(t) is a piecewise continuous function which is integrable on any finite subinterval for $t > 0$, a hybrid differentiable-integral operator is given by,

$$_0\partial_t^{-v} f(t) = \frac{1}{\Gamma(v)} \int_0^t (t - \xi)^{v-1} f(\xi)d\xi \quad _0 d_t^{-v} f(t) = \frac{1}{\Gamma(v)} \int_0^t (t - \xi)^{v-1} f(\xi)\partial\xi$$

$$_a\partial_t^0 f(t) = f(t) \quad and \quad _0\partial_t^1 f(t) = \left. \frac{df(t)}{dt} \right|_{backward} \sim limit_{\Delta t \to 0} \frac{n(t) - n(t - \Delta t)}{\Delta t}$$

Generalizing the differ-integral kinetics equation obtains,

$$-c \cdot n(t) = {_0\partial_t^{-v}} n(t)$$

Taking Laplace Transforms, $-c \cdot \bar{n}(s) = s^v \bar{n}(s) - s^{v-1} n(0^+)$

and therefore, $\qquad \bar{n}(s) = \frac{n_0 + s^{v-1}}{(s^v + c)} = \frac{n_{0+}}{s^{1-v}(s^v + c)}.$

Inverting the transform,

$$L^{-1}\left[\frac{1}{s^\alpha(s^\beta - a)}\right] = \sum_{j=1}^w a^{j-1} E_t\,(j\beta + \alpha - 1, a^w)$$

$$n(t) = n_{0+} \sum_{j=1}^w (-c)^{j-1} E_t\,(v(j-1), (-c)^w)$$

The modified Mittag-Leffter function is denoted by, $\qquad E_t(v, a) = t^v \sum_{k=0}^\infty \frac{(at)^k}{\Gamma(v+k+1)}$

introduced as a generalization of the exponential in solving linear ordinary differential equations.

Further fractional calculus representations include,

$$\frac{\partial^q f}{\partial(x-a)^q} = limit_{N \to \infty} \left\{ \frac{h^{-q}}{\Gamma(-q)} \sum_{j=0}^{N-1} \frac{\Gamma(j-q)}{\Gamma(j+1)} f(x - jh) \right\}$$

Where $q = -1,$ Riemann sum limit,

T. Margulies, *Pyramid Geometry Design*

$$\frac{\partial^{-1}f}{\partial(x-a)^{-1}} = limit_{N\to\infty}\left\{h\sum_{j=0}^{N-1}f(x-jh)\right\}$$

$q = 1,$ Backward difference sum limit,

$$\frac{\partial^{1}f}{\partial(x-a)^{1}} = limit_{N\to\infty}\left\{\frac{1}{h}\sum_{j=0}^{N-1}f(x-jh)\right\}$$

Modified improved convergence Grunwald differ-integral difference formula,

$$\frac{\partial^{q}f}{\partial(x-a)^{q}} = limit_{N\to\infty}\left\{\frac{1}{h^{q}\Gamma(-q)}\sum_{j=0}^{N}\frac{\Gamma(j-q)}{\Gamma(j+1)}f(x-(j-0.5q)h)\right\}, \ h = \frac{x-a}{N}$$

Also, the Mittag-Leffler functions have alternate mathematical representations, such as, the incomplete gamma function, $\gamma^{\star}$, hypergeometric series, $F,$ and others.

$$E_{t}(v,a) = t^{v}e^{at}\gamma^{\star}(v,at) \quad \text{or} \quad E_{t}(v,a) = \frac{t^{v}}{\Gamma(v+1)} \ _{1}F_{1}\ (1,v+1;at)$$

where $$\gamma^{\star}(v,at) = \int_{0}^{t}y^{v-1}(1-y)^{at-1}\ dy \quad \text{and}$$

$$_{p}F_{q}\ (a_{1.}\dots a_{p}, b_{1.}\dots b_{q}; t) = \frac{\Gamma(b_{1.})\cdots\Gamma(b_{q.})}{\Gamma(a_{1.})\cdots\Gamma(a_{p.})}\sum_{k=0}^{\infty}\frac{\Gamma(a_{1.}+k)\cdots\Gamma(a_{p}+k)\ t^{k}}{\Gamma(b_{1.}+k)\cdots\Gamma(b_{q}+k)\ k!}$$

Reference: Miller, K. S. and B. Ross, *An Introduction to Fractional Calculus*, John Wiley & Sons (1993).

T. Margulies, *Pyramid Geometry Design*

www.ingramcontent.com/pod-product-compliance
Lightning Source LLC
Chambersburg PA
CBHW050733180526
45159CB00003B/1209